THE GENERATION AND TRANSMISSION OF ELECTRICITY: A HISTORICAL VIEW

by

DR. Peta Trigger Ph.D, Ed.D (Lon)
*N*orthampton *A*cademy of *P*ostdoctoral *S*tudies
NN3 8TJ

First published 2013. Revised version published in the UK by *NAPS* © 2020

ISBN 9781494319342

Printed in the UK

FOREWORD

This book is based on Part I of the author's post-doctoral dissertation investigating two electricity topics taught at undergraduate level. It contains experimental and research notes and a number of figures, some in the form of sketches, including illustrations, drawings of machines, circuit diagrams, tables and graphs of experimental data, together with their annotations, which were produced at the same time as the experimental work and background research were being carried out. With the aim of authenticity, to preserve these figures as far as possible in their original form, they were not completely re-drawn but tidied and then photocopied for inclusion in this book. Also, whilst in some instances, annotations have been expanded and captions added for the purpose of clarification, the content of the original annotations has been retained in its original form. The author takes the view that this adds authenticity to the book to an extent which justifies any slight loss in the 'professional' look of the book which might have resulted from re-drawing and typing. However, where required, additional sub-headings to improve readability have been included, and where necessary, corrections to the text and calculations have been made.

TABLE OF CONTENTS

Preface   17

1. INTRODUCTION   19
THE ANCIENT ELECTRICIANS

2. DR.GILBERT
DE MAGNETE   30

3. ELECTROSTATIC ELECTRICITY GENERATION

THE ELECTROSCOPE   35

THE ELECTROPHORUS   38

EARLY ELECTROSTATIC GENERATING MACHINES   39

VAN DER GRAPH GENERATOR   52

THE DISCOVERY OF ELECTRICAL
CONDUCTION   57

THE LEYDEN JAR   58

4. ELECTROCHEMICAL ELECTRICITY GENERATION

GALVANI AND 'GALVANISM'   68

VOLTA   70

THE VOLTAIC PILE   72

5. ELECTROMAGNETIC ELECTRICITY GENERATION

OERSTED   81

FARADAY   88

FARADAY'S DISC GENERATOR   96

PIXII MAGNETO-ELECTRIC GENERATOR	101
THE EVOLUTION OF THE ARMATURE	107
SEPARATELY EXCITED VS SELF-EXCITED GENERATING MACHINES	113

6. ELECTRIC LIGHTING

THE NEED FOR SUPPLY	115
THE ARC LAMP	115
CROMPTON'S LIGHTING SYSTEM	120
THE INCANDESCENT LAMP	125
EDISON'S LIGHTING SYSTEM	136
FERRANTI	145
THE BATTLE OF THE SYSTEMS	157

6

THE IMPORTANCE OF THE STEAM TURBINE 164

7. THE TRANSMISSION OF ELECTRICITY

EARLY CABLES 174

EDISON'S DISTRIBUTION SYSTEMS 185

EARLY H.T. CABLES 188

3-PHASE A.C. 198

SOME PROBLEMS IN E.H.T. CABLE DESIGN 201

8. CONCLUSION 207

9. BIBLIOGRAPHY 221

10. APPENDICES 224

ABOUT THE AUTHOR 290

LIST OF FIGURES

FIG.1 THE PRINCIPLE OF THE ELECTROPHORUS   38

FIG. 2   THE PRINCIPLE OF THE VAN DER GRAAF GENERATOR   55

FIG. 3 COMMUTATING ALTERNATING CURRENT: 104

FIG. 4   CROMPTON-BURGIN ARMATURE   110

FIG. 5 CROMPTON-BURGIN GENERATOR   111

FIG 6 . CROMPTON ARC LAMP   119

FIG. 7   . PORTABLE GENERATOR   122

FIG. 8. CROMPTON'S ARC-LIGHTING AT ALEXANDRA PALACE   123

FIG. 9. FERRANT'S ZIG-ZAG ARMATURE   146

FIG. 10 FERRANTI-THOMPSON DYNAMO   147

FIG. 11 FERRANTI TRANSFORMER AS USED AT DEPTFORD/GROSVENOR SQUARE   149

FIG. 12 FERRANTI 750 H.P. ALTERNATORS USED AT GROSVENOR GALLERY   152

FIG 13 DEPTFORD POWER STATION   153

FIG. 14 FERRANTI 10 kV TRANSFORMER   155

FIG. 15 FERRANTI HIGH VOLTAGE DISTRIBUTION SYSTEM   156

FIG. 16 FERRANTI RECTIFIER   158

FIG. 17 FERRANTI STEAM TURBO-GENERATORS   172

FIG. 18 EDISON'S THREE WIRE SYSTEM   186

FIG. 19 ARRANGEMENT OF COILS IN A THREE-PHASE GENERATOR   199

FIG. 20 A 3-PHASE SYSTEM WITH BALANCED OR SYMMETRICAL LOADS   200

FIG. 21 GRAPH SHOWING HIGHEST SYSTEM VOLTAGES USED IN BRITAIN SINCE 1891   218

FIG. 22 GRAPH SHOWING THE INCREASE IN DEMAND FOR ELECTRICITY SINCE 1920   219

FIG. 23 A REPLICA VAN DER GRAAF GENERATOR   224

FIG. 24 A REPLICA LEYDEN JAR   229

FIG. 25 CIRCUIT USED TO DETERMINE CAPACITANCE OF A LEYDEN JAR   232

FIG. 26 DETERMINATION OF MAXIMUM VOLTAGE OF A LEYDEN JAR   234

FIG. 27 IONIZATION OF A METAL IN WATER AND THE PRODUCTION OF A POTENTIAL DIFFERENCE   239

FIG. 28 A REPLICA VOLTAIC PILE   240

FIG. 29 THE CIRCUIT USED TO MEASURE BATTERY PERFORMANCE   244

FIG 30 VARIATION IN P.D. OF A VOLTAIC PILE WITH COMPRESSIVE FORCE ON PLATES   246

FIG. 31 PERFORMANCE OF A VOLTAIC PILE (3Ω LOAD) : ONE CELL (PAPER TOWELLING SEPARATOR)   249

FIG. 32 VOLTAIC PILE: GRAPH OF PERFORMANCE WITH 3Ω LOAD   250

FIG. 33 GRAPH OF VOLTAIC PILE CURRENT VS TIME OF DISCHARGE   252

FIG. 34 CIRCUIT USED TO COMPARE THE PERFORMANCE OF A VOLTAIC PILE WITH A MODERN LECLANCHE CELL   253

FIG. 35 GRAPH OF LECLANCHE CELL PERFORMANCE INTO A 3 OHM LOAD   257

FIG. 36 REPLICA FARADAY DISC GENERATOR: SIDE VIEW   259

FIG. 37 A REPLICA FARADAY DISC GENERATOR: TOP VIEW   260

FIG. 38 CIRCUIT FOR PERFORMANCE TESTING OF A FARADAY DISC GENERATOR   261

FIG. 39 GRAPH OF FARADAY DISC GENERATOR CURRENT VS LOAD   263

FIG. 40 GRAPH OF FARADAY DISC GENERATOR CURRENT VS 1/LOAD   264

FIG. 41 A REPLICA PIXII GENERATOR   272

FIG. 42 GRAPH OF CURRENT VS LOAD OF A PIXII GENERATOR   277

FIG. 43 DEMAGNETISING COMPONENT ON THE ARMATURE MAGNETIC FIELD DUE TO COMMUTATION   279

FIG. 44 GRAPH OF OUTPUT CHARACTERISTIC OF THE SEPARATELY EXCITED GENERATOR   278

FIG. 45 GRAPH OF OUTPUT CHARACTERISTIC OF A SHUNT WOUND GENERATOR   282

FIG 46 GRAPH OF OUTPUT CHARACTERISTIC OF A SERIES WOUND GENERATOR   284

FIG. 47 GRAPH OF OUTPUT OF A COMPOUND WOUND GENERATOR   286

FIG 48 THE THREE PHASE VOLTAGES IN A THREE-PHASE GENERATOR   287

FIG. 49 VECTORIAL REPRESENTATION OF THE THREE PHASE VOLTAGES IN A THREE PHASE SYSTEM   288

LIST OF TABLES

TABLE 1 SHOWING THE RELATIONSHIP BETWEEN THE LENGTH OF SPARK IN AIR BETWEEN SPHERES OF VARIOUS DIAMETERS   228

TABLE 2 CALCULATION OF INTERNAL RESISTANCE: FILTER PAPER SEPARATOR   243

TABLE 3 CALCULATION OF INTERNAL RESISTANCE: PAPER TOWELLING SEPARATOR   243

TABLE 4 THE VARIATION IN P.D. OF THE VOLTAIC PILE WITH COMPRESSIVE FORCE ON THE PLATES
                                                                           245

TABLE 5 PERFORMANCE OF A VOLTAIC PILE: ONE CELL   248

TABLE 6 PERFORMANCE OF ONE VOLTAIC CELL WITH FILTER PAPER SEPARATOR INTO 3Ω   249

TABLE 6 VOLTAIC PILE: PERFORMANCE WITH 1 000 Ω LOAD   248

TABLE 7 VOLTAIC PILE PERFORMANCE WITH 1 000 Ω LOAD   251

TABLE 8 LECLANCHE CELL INTERNAL RESISTANCE 256

TABLE 9 FARADAY DISC GENERATOR PERFORMANCE 262

TABLE 10 HIPPOLYTE PIXII GENERATOR CURRENT VS LOAD 276

LIST OF PLATES

PLATE 1. THE ELECTROSCOPE   37

PLATE 2. THE GLOBE ELECTRIC GENERATOR   41

PLATE 3. THE CYCLINDER ELECTRIC MACHINE   47

PLATE 4. THE PLATE ELECTRIC MACHINE   49

PLATE 5. THE WIMSHURST ELECTRIC MACHINE   51

PLATE 6. VAN DER GRAAF GENERATOR   54

PLATE 7. THE LEYDEN JAR   58

PLATE 8. VOLTA'S 'PILE'   76

PLATE 9. FARADAY'S DISC GENERATOR   97

PLATE 10 . PIXII GENERATOR   102

PLATE 11. EARLY GRAMME DYNAMO 106

PLATE 12. VARLEY'S SELF-EXCITED
GENERATOR   114

PLATE 13. EDISON BIPOLAR DYNAMO
                                     143

PLATE 14. FERRANTI HIGH VOLTAGE
CABLES   192

PREFACE

This book is concerned with the history of the generation and transmission of electricity from the ancient Greeks to modern times. Special emphasis is given to those generating machines and other electricity detecting and storage devices which, arguably, represent landmarks in development. Amongst these are the Leyden jar, which, as a device for storing electrostatic electricity was as important to development of the electrostatic generator as the voltaic pile battery, storing current electricity, was to the electromagnetic generator; the van der Graaf machine which represents a culmination in the application of the main principles of electrostatics discovered in the development of electrostatic generators; and the magneto-electric Faraday Disc and Pixii generators which were the first machines to incorporate the principle of electromagnetic induction on which all subsequent generators of current electricity are based.

The author made working models of each of these devices, and the book contains details of their construction and experiments to test their performance, and where appropriate compare them with modern counterparts.

The book contains many diagrams, tables and graphs to these ends and also included are copies of photographs and drawings of historically significant electricity detectors, electrostatic generators, electrostatic and battery storage devices, electromagnetic generators and components, electric light and lighting systems, and cables and distribution systems, together with some illustrations of their use and application in practice.

1. INTRODUCTION

THE GREEKS AND THE ANCIENTS

Early Electrostatics and the Properties of Amber

The word 'electricity' comes from 'electron', the Greek word for amber. Amber resembles gold in its polished yellow brilliance and was used as barter between wealthy tradesmen from the East and Ancient Greece. Thales of Miletus (640 - 546 B.C.) observed that amber could be 'quickened' by rubbing with a cloth and it would attract small light objects such as pieces of fabric and leaves. It 'possessed a soul of its own' and was held sacred. It was used in making bracelets and necklaces which gave off a pleasant odour when rubbed.

Three centuries later Theophrastus discovered that other materials became electrified by rubbing. However the Greeks did not seem aware that it was the same force that causes thunder and lightning but explained these phenomena in terms of supernatural creatures fighting battles in the sky, or gods showing their anger.

Greek women decorated their spinning wheels with pieces of amber. A modern explanation is that as the woollen threads rubbed against the amber it first attracted and then repelled them. As an opposite charge was induced on the fabric it was attracted to the amber. On touching it the fabric acquired some of the amber's charge and was repelled by it- a pretty little spectacle which relieved the boredom of spinning.

Electrostatic Phenomena in Biblical Times

Perhaps the Israelites knew something about electricity. They had metal rods on the roof of the Temple at Jerusalem which may have functioned as lightning conductors. Indeed it was never struck by lightning in all its history even though thunderstorms are frequent in Jerusalem.

It is possible that Aaron's sons as described in Leviticus 10:2 were electrocuted by the discharge from the sacred shrine which contained the tablets with the Ten Commandments written on them. The container was made from acacia wood lined inside and out with thin sheets of gold. This therefore acted as a capacitor on which large amounts of electrical charge and a high voltage might have built up from atmospheric electricity communicated to the acacia wood 'dielectric' by the large quantities of precious metals stored with it.

The Discovery of Properties of Magnets and Lodestones and Early Attempts to Explain Them

Reports from the ancients on the subject of magnetism are much more forthcoming than electrostatics, one reason being that properties of magnets seem more spectacular and the forces involved are greater. Also, it was not necessary to rub magnets to make them perform; they possessed a 'magic' of their own.

Early Greek manuscripts tell of small pieces of ore found by Phrygian miners, the Cabiri, whose wondrous properties were demonstrated by a travelling band of magicians at Samothrace. The ore could support a column of rings one below another without them being fixed together. Plato described the Samothracian rings in a letter to Ion dating from the 4th century B.C., and they are also described in other early Greek literature:

> *There is a divinity moving you like that contained in the stone which Euripides calls a magnet...This stone not only attracts iron rings, but also imparts to them a similar*

> *power of attracting other rings...and all of them derive their power of suspension from the original stone in like manner as the Muse first of all inspects men herself and from these persons a chain of other persons is suspended who take the inspiration.*

The Cabiri were transformed into legend as dwarfs who spent their time excavating the magical stone from the earth.

There are numerous accounts of the origin of the word 'magnet'. Pliny claims that it arose from the discovery of lodestone by a shepherd named Magnes. Lecretius wrote that lodestone was first discovered in Magnesia from which 'magnet' arose. It is not surprising that the ancient Greeks viewed the properties of magnets as remarkable, especially the force of attraction between magnet and object separated by air. Since the ancients could not conceive of a physical law of magnetic attraction, magnets were esteemed divine objects having much in common with the spiritual world.

The ancients were quick to try the effect of lodestone on a variety of metals and non-metals but discovered that the magnet's affinity seemed only for iron. This is clear from the name given to magnets by the early Chinese.

Early Attempts to Explain Magnetism

The first philosopher to give serious thought to the nature of magnetic force was Thales of Miletus, an ancient city in Asia Minor. His writings are lost, but one of his commentators, Aristotle, hypothesised that:

> *The magnet has a soul because it causes movement to iron.*

Further searching for a mundane explanation of magnetic attraction was held in check at this time because of the character of the religion of the ancients and the firm belief that the powers of magnets were associated with the gods and any attempts at investigation would anger them. Much later, some ventured to attempt to explain magnetic attraction in what seemed to them to be 'natural' terms, and although these explanations are fantastic, this epoch is significant because it marks the beginning of a long search for a correct physical explanation of magnetism. Epicurus (342-

270 B.C.) wrote:

> *The lodestone attracts iron because the particles which are continually flowing from it, as from all bodies, have such a peculiar fitness of form that upon collision they easily unite.*

Lucretius (99 -55 B.C.) wrote:

> *Moreover, it is possible that some things are held together linked and interwoven as though by rings and hooks; which seems more likely to be what happens with the iron and stone.*

Plutarch (46-120 A.D.) imagined the pores of iron to have a shape similar to the particle of an assumed aura surrounding the magnet and thus iron caught and held a magnet in a way which no other substance could. Claudian (300 A.D.) reported:

> *Iron gave life to the magnet and nourished it, hence the magnet sought iron as animals sought food.*

Pliny explained magnetic attraction as follows:

> *The moment the metal comes near it, it springs toward the magnet and as it clasps it is held fast in the magnet's embrace*

The Discovery of Residual Magnetism and Magnetic Repulsion

The ancient Greeks discovered that the residual magnetism in iron brought into association with a lodestone in certain circumstances led to repulsion.

> *There are times also when the nature of iron moves away from this stone, being wont to flee and follow it by turn. I have seen rings of Samothracian iron dance, and iron filings leep about inside brass bowls, soon as this magnet stone is put beneath, so eager the iron seems to escape from the stone.*

Pliny called the stone exhibiting these repulsion properties 'theamides', a different type of stone to lodestone found in Ethiopia:

> *The leading distinction in magnets is sex, male and female...The kind that is found in Treas is black of the female sex and consequently destitute of attractive power.*

The Discovery of the Magnetic Compass Needle

The ancients knew that iron objects could be magnetised by residual magnetism brought about by association with lodestone, but the application of this phenomenon to produce a compass needle is not recorded until a 121 A.D. when the Chinese explained the action of a needle magnetised by lodestone pointing south. It seems probable that the Chinese rested the magnetic needle on the surface of a bucket of water whence it aligned itself in the magnetic meridian. The significance of this discovery for navigation was not appreciated until many centuries later.

The Beginnings of Experimental Method in Investigating the Properties of Magnets and New Discoveries

The early experimenters lacked correct experimental method- their observations were inaccurate and their hypotheses fantastic and unproven. Little serious interest was taken in magnetism until the late 12th century when reference to the use of a magnet as a mariner's compass appears in the literature of that time. The revival in interest in magnetism after this time was due in no small way to the efforts of Roger Bacon in the late 13th century. In his work *Opus Minus* he records his observations of magnetic repulsion and points the way to the discovery of the concept of magnetic polarity.

The most important work of this time was carried out by Peter Peregrine in 1269 in response to an inquiry about magnets from a close friend. Peter Peregrine was also a close friend of Roger Bacon whose work no doubt stimulated further interest within his circle of scientific friends. Peregrine's work 'Epistola de Magnetica' consisted of thirteen chapters in which are recorded careful observations, hypotheses and verification of hypotheses by further experiments on magnetic polarity, and on repulsion and the concept of line of force.

However, the strongest influence to re-examine the properties of magnets was brought about by the inconsistent behaviour of magnetic compass needles at different places on the Earth's surface. Early Chinese mariners found that their compasses did not point accurately north but pointed at an angle to the north-south line dependent on their position. In fact a wide range of angles both to the east and west of north were found in different locations. On the epic voyages of Vasco da Gama and Christopher Columbus declination or 'dip' of the compass needle was also observed.

A discovery by Georg Hartmann of Nuremburg showed that magnetic needles were deflected vertically as well as horizontally, and this was later brought to the attention of Robert Hartmann who drew up a chart of magnetic declination in 1580 and published findings of magnetic dip in his *'Newe Attractive'* in 1581.

These discoveries were of such importance that they demanded the attention of scientists in achieving some insight into their mechanisms. This led the Elizabethan scientist Gilbert to propound a theory of terrestrial magnetism in his treatise *'De Magnete'*.

Having summarised the work of the ancient electricians in this introduction, the work of scientists, beginning with Gilbert, which led to development of the first electrostatic generating machines, and then Volta's work on current electricity and electrochemical batteries, followed by the discoveries by Oersted and Faraday of electromagnetism and electromagnetic induction, which ultimately led, via Parson's work on steam turbines, to the development of modern steam-driven electromagnetic generators and modern methods of electricity distribution, will be reviewed. Several replica generators, a 'condenser' and a battery, modelling the devices which were landmarks in the history of the generation and transmission of electricity, will also be described and their performance investigated.

2. DR. GILBERT
DE MAGNETE

In Gilbert's famous treatise *'De Magnete'* written in 1600 was contained all the knowledge accumulated about electricity and magnetism susceptible to verification and experiment up to that time, together with many innovations. This work was a great contribution to science and knowledge in the very essence of the term Renaissance, for it revived interest in electricity and magnetism which had made little progress since Ancient Greek times. It brought the sum total of knowledge of the subject under one cover for all interested parties to review as a foundation for further experimental investigation.

THE DISCOVERY OF MAGNETIC POLES AND THE EARTH'S MAGNETISM

The mariners' compass was first heard of about 1200 but the people of the time were unsure of the explanation for its behaviour. To the medieval mind, the answer was that the compass needle pointed at the pole star. However, Columbus in his voyages in the Atlantic in the later 15th century found that it did not point to true north but a few degrees from it. In 1544 it was discovered that a compass needle, pivoted so that it could move in a vertical plane, did not remain parallel to the Earth's surface but dipped down towards it. With these facts in his possession, Dr. Gilbert introduced the idea that the Earth was a magnet. To test this idea, he constructed a small Earth or globe out of a piece of lodestone. This he called a terrella or 'earthlet'. He placed a series of iron wires on the terrella and observed the angles of dip in various latitudes. He found that the positions taken up by the iron wires were analogous to the angles of dip and declination recorded by Columbus during his voyage to the East Indies. Not surprisingly, he also discovered that a compass needle pointed north because it was attracted to the Earth's magnetic north pole. The 'North Pole' was the point to which the 'lines' of iron filings pointed when they were sprinkled in the vicinity of a piece of lodestone. There was another pole at the opposite end of magnets to which Gilbert gave the name 'South Pole'. Gilbert found the density of lines in the region between the poles to be less than near the poles, not attracting the iron filings so strongly. Half way between the poles, which he called the equator, iron filings were not attracted at all.

THE DISCOVERY OF THE LAW OF ATTRACTION AND REPULSION BETWEEN MAGNETS

Gilbert suggested that the force of magnet might be measured by making it attract a piece of iron attached to one arm of a balance, weight being added to the other until the magnet 'let go'. He distinguished between 'magnets' and 'magnetic substances'. A magnet attracts another magnet if unlike poles are brought together and like poles repel. But a lump of iron is attracted to either pole of a magnet and has itself no distinguishable magnetic poles or equator, no matter which part was brought close to a magnet.

INDUCED MAGNETISM

Some of Gilbert's experiments were concerned with investigating circumstances in which a magnet could induce magnetism in iron. He placed different materials such as glass, wood and paper between a piece of iron and a lodestone and found that the iron was always influenced by the magnet whatever the intervening material- except iron. He even tried heating a piece of iron in a flame whilst a magnet was held close, but the magnet still attracted the iron. However, he found that magnets were destroyed by strong heating and banging.

Gilbert utilised the Earth's magnetism to magnetize a bar of steel. He placed the bar in the magnetic meridian and found that it became weakly magnetized. He also states that iron bars left upright for a long time acquire magnetism from the Earth, and banging a bar of steel in the magnetic meridian produces a magnet very quickly. If a bar of steel is heated to redness whilst lying the magnetic meridian it becomes magnetized when allowed to cool in that position, whilst no such effect occurs if it is left pointing in an East-West direction.

Gilbert endeavoured to demonstrate the truth of his conclusions and those of others before him by experiment. In his treatment of amber and lodestone he writes:

> *Great has ever been the fame of lodestone and of amber in the writings of the learned; many philosophers cite the lodestone and also amber whenever, in explaining mysteries, their minds become obfuscated and reason can no further go. Over-inquisitive theologians, too, seek to light up God's mysteries and things beyond man's understanding by means of lodestone and*

> *amber... Thus in very many affairs, persons who plead for a cause the merits of which they cannot set forth, bring in as masked advocates the lodestone and amber. But all these, besides sharing the general misapprehension are ignorant that the causes of the lodestone's movement are very different from those which give to amber its properties; hence they easily fall into errors and by their imaginings are led farther away.*

So it was clear that Gilbert dismissed the myths about magnetism and electricity, and in *De Magnete* he describes many experiments in which his theories are proven. In his careful cataloguing of his work he invited subsequent experimenters such as Gray and von Guericke to follow his reasoning and extend his findings, leading to more discoveries in electrostatics and magnetism. The work of Gray and von Guericke is described in the next chapter.

3. ELECTROSTATIC ELECTRICITY GENERATION

GILBERT'S WORK AND LATER DEVELOPMENTS

THE ELECTROSCOPE

In his work with 'electrics' - the 'things which attract as amber does', Gilbert constructed the first recorded crude electroscope or 'vasorium' made of

> ...*any sort of metal, three or four fingers long, pretty light, and poised on a sharp point after the manner of a magnetic pointer.*

When he rubbed amber and brought it near to either end of his needle, it turned towards the amber. In the same way, he examined many other materials which he found behaved in the same way:

> *It is not a rare property possessed by one or two (as is commonly supposed) but evidently belongs to a multitude of objects, both simple and compound... For not only do amber and jet, as they suppose, attract light substances, the same is done by diamond, sapphire, carbunkle, iris stone, opal, amethyst, vincentina, Bristol stone, beryl and rock crystal.*

The foregoing materials he termed 'electrics'. He also compiled a list of 'non-electrics', including metals, which would not attract the end of his electroscope because any charge which was resident on them flowed to or was neutralised by earthing, though of course Gilbert did not realise this. During his experiments conducted in foggy, damp London he found that they did not succeed in humid conditions.

THE IMPORTANCE OF THE ELECTROSCOPE IN EXPERIMENTS

The Renaissance work of Gilbert was a scientific study which stimulated new interest in magnetism and electrostatics. The concept of magnetic poles, attempts to measure magnetic field density and magnetic induction, were all subjects of his experiments which stimulated later work with magnets which ultimately led to the discovery of electromagnetic induction. His work on electrostatics was important too, particularly the electroscope which in later developments was a tool for experimenters to use in investigating electrostatics phenomena:

Fig. 1-1. The leaf electroscope

Plate 1. THE ELECTROSCOPE. *Source: 'Electricity and Magnetism'*

Gilbert had opened up a field of study for many new experiments in electrostatics.

THE ELECTOPHORUS

The electrophorus was devised in 1775 by Alessandro Volta. By means of this device, he found that it was possible to convey an 'unlimited' number of charges from one single charge:

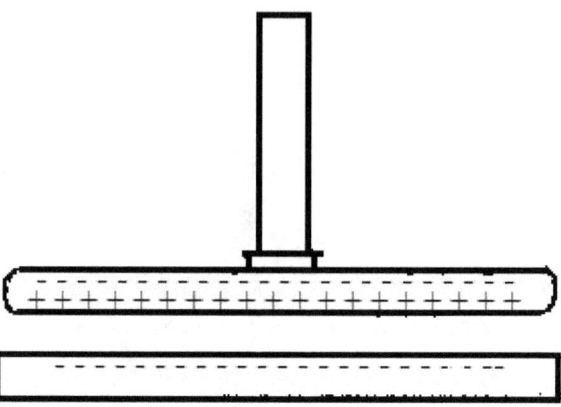

FIG. 1. THE PRINCIPLE OF THE ELECTROPHORUS

It consisted of two parts, a round cake of resinous material cast into a metal dish or 'sole' about 12 inches in diameter, and a second round disc of slightly smaller diameter made of metal provided with a glass handle (see above). Shellac or sealing wax was used to make the 'cake'. To use the electrophorus, it had to be rubbed with a warm dry cloth or catskin. The disc was then placed on the cake, momentarily touched with a finger, then removed by taking it up with the glass handle.

By this means, the disc acquired a positive charge of such high potential that a spark jumped from the disc to a finger placed near it. The disc could then be replaced, touched, removed and once more became charged by induction, a process which could be repeated many times. In modern terms, when the resinous cake is first rubbed with the catskin its surface becomes negatively charged by acquiring an excess of electrons. When the metal disc is placed upon it, it rests only on three or four points and is therefore virtually insulated from the cake. The negative charge on the cake induces a positive charge on the metal disc, and a finger placed on it provides a conducting path for the electrons to earth. The charge on the disc then spreads out and it acquires a net positive charge.

EARLY ELECTROSTATIC GENERATING MACHINES

Following Gilbert, one man, Otto von Guericke (1602-1686), decided that electrostatics phenomena could be investigated more effectively if a machine to produce 'electric virtue' for experiments could be devised. von Guericke's electrostatic machines, which led to his discovery of electrical conduction, were also of paramount importance in the evolution of the electricity generator, for they were examples of the electrostatic principle in machines which magnified the effect many times and produced continuous electricity (though in very small quantities): an analogous electromagnetic principle was to be used in machines which magnified the effect of a wire cutting the lines of a magnetic field to produce current electricity. However, electromagnetic induction and the generation of measurable quantities of electricity still lay far in the future.

VON GUERICKE'S GLOBE ELECTRIC GENERATOR

To save himself the labour of rubbing objects to produce electricity for his experiments, he decided to make a machine to do it for him :

PLATE 2. THE GLOBE ELECTRIC GENERATOR. *Source: Science Museum*

Von Guericke used sulphur to make the globe because it was supposed to share the same 'infernal' nature as sulphur and was a favourite electric of the early experimenters. von Guericke poured melted sulphur into a glass globe, breaking the glass when the sulphur had solidified. It was supplied with a wooden handle and mounted in a frame in which it could be rotated. A similar globe and handle, on a smaller scale, was used to withdraw some of the 'electric virtue' from the larger globe which was charged by resting the palm of the hand on it whilst it was rotating.

THE ROTATING DISC ELECTRIC FRICTION GENERATOR

von Guericke is also credited with a more advanced machine consisting of a metal disc rotated between two brushes which pressed against it. Friction between the brushes and the disc induced opposite charges on the brushes and disc. The charge from the brushes was collected on a spherical conductor. Although von Guericke had built machines which were used for countless experiments, they were otherwise little more than a form of entertainment in polite society in the late

17th century. von Guericke did not realise that the glass phial he used to make his sulphur globe would have served as well or better.

Attempts to Explain Electrical Phenomena

It was Hawksbee in 1709 who discovered the effectiveness of glass as a means of producing electricity. He made an electrical machine consisting of a glass globe rotated against a rubber of coarse woollen cloth. Up to the early 1730's, scientists found the phenomenon of electricity unaccountable and propounded no satisfactory laws to describe it. Gilbert and Hawksbee invented the doctrine of 'effluvia' which was likened to a vapour given off by bodies heated by rubbing just as steam is given off by hot water. Thus during this time 'electric virtue' was seen as part of the body whose effects are observed rather than a phenomenon capable of existing independently. Electric virtue was some 'fluid' which a limited number of substances contained and was 'exuded out' by friction.

In the 1730's Stephen Gray experimented with the conduction of electricity and made an attempt to classify materials in terms of 'electrics' and 'non-electrics', but throughout his work he makes no *scientific* attempt at explaining electricity, clinging to the old concept of 'effluvia'. Desagulier, a contemporary of Gray, employed the concept of two different kinds of electricity discovered by DuFray to explain attraction and repulsion. He wrote about a sphere of effluvia around electrified objects after the manner of the effluvia of sulphur which spread out in a room combining with particles of air. It is possible that he viewed effluvia as consisting of repelling particles forming a sphere of influence around electrified objects. He says the effluvia are 'thrown out' of glass tubes by rubbing , thereby producing one of two kinds of electricity in the glass. When it communicates its electricity to another body it loses its electricity 'at once' and as it does so 'the effluvia coming out strike the new body'.

William Watson in the 1740's remarked upon the luminous discharge of a sword connected to an electrostatic machine and used the term 'electric fire' instead of 'electric virtue' to describe it. Watson discovered that 'electrics' were not a source of electricity themselves but that electricity was resident on them. By insulating his electrostatic machine and its operator, he found that he could not induce sparks from the machine and concluded that electricity 'came up from the floor'. Finally, Watson discovered the principle of operation of the early frictional machines:

> *The glass globes circulate the electrical fire which they receive from their friction against the cushions or the hand of a man, and which is constantly supplied to these last from the floor.*

The way was cleared for efficient high voltage electrostatic machines when Watson's contemporary Benjamin Franklin discovered positive and negative electricity and the discharging effect of points.

IMPROVEMENTS TO VON GUERICKE'S GLOBE GENERATOR

The von Guericke machine was improved chiefly by German scientists, one of whose innovations was to provide a handle containing a large wheel which communicated with a much smaller wheel attached to the whirling globe via a length of cord.

Watson estimated that that the machine he used in 1746 carried a globe which was rotated at a mean speed of 1100 r.p.m.. To improve contact between the 'rubber' or friction pad and the rotating globe, a tin amalgam was later used to coat the leather pad and globe. For use with his experiments, Watson added a gun barrel to his generator which was connected to the globe by means of 'metallic fringes' which hung down onto it. After Franklin's discovery of the discharging effects of points, a metallic comb was used to 'collect' the charge from the globe and deposit it on to the gun barrel, thenceforth called the conductor.

THE CYLINDER ELECTRIC MACHINE

PLATE 3: CYLINDER ELECTRIC MACHINE.
Source: Science Museum

The fully developed von Guericke machine, containing all the innovations mentioned was called the Cylinder Electric Machine (see above), consisted of a glass cylinder mounted horizontally and was turned by a handle. A leather cushion stuffed with horse hair, the surface of which was coated with an amalgam of tin or zinc, was pressed against it from behind. A flap of silk attached to the cushion passed over the cylinder, covering its upper half. In front of the cylinder stood the 'prime conductor', which was made of metal usually in the form of an elongated cylinder with hemi-spherical ends, mounted on a glass stand.

At the end of the prime conductor nearest the cylinder axis was a fixed rod bearing a row of fine metallic spikes, resembling a rake. The other end usually carried a rod terminated in a brass ball or knob. When the handle is turned, the friction between the glass and the amalgam-coated rubber produces a positive charge on the glass and a negative charge on the rubber. The prime conductor, via the spikes, sprays a negative current onto the cylinder to neutralise its positive charge. At each revolution of the cylinder, therefore, an increment of positive charge is left on the prime conductor. A metallic knob at the back collects the negative charge.

THE PLATE ELECTRIC MACHINE

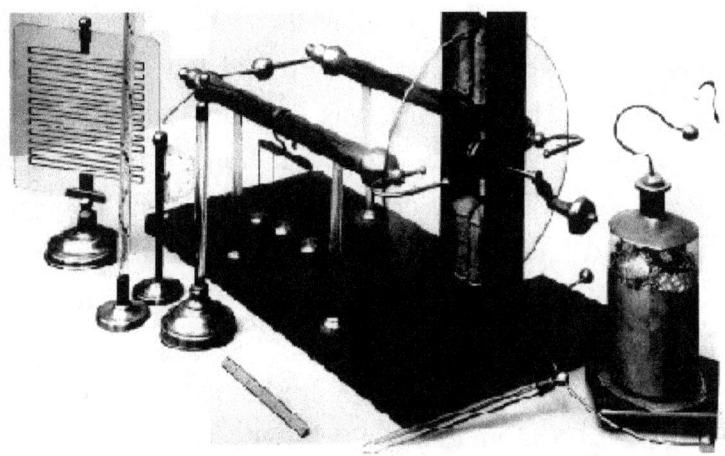

PLATE 4. THE PLATE ELECTRIC MACHINE.
Source: Science Museum

A further development of the Cylinder Electric Machine was the Plate Electric Machine in which a disc replaced the original rotating cylinder (see above). This machine had the advantages over the Cylinder Machine that the large disc was easier to make and the plate machine was capable of generating a much higher voltage because, due to the large size of the disc, the discharge path of the machine was longer.

The Plate Electric Machine consisted of a circular glass or ebonite plate and was usually provided with two pairs of rubbers in the form of double cushions, pressing the plate between them. These were placed high up and low down on the plate. The machine was provided with silk flaps in the shape of a circle quadrant. The prime conductor was either double or curved round to meet the plate at the two ends of its horizontal diameter and was furnished with two sets of points or spikes which 'collected' the charge as in the Cylinder Electric Machine. The Wimshurst machine was a later variant (shown below):

51

PLATE 5. THE WIMSHURST ELECTRIC MACHINE. *Source: Science Museum*

Shortly after 1775 and Volta's discovery of the electrophorus for transferring charge, Nicholson, Bennet and others designed a revolving apparatus capable of utilizing Volta's concept in a generating machine. Nicholson's revolving doubler, invented in 1788, consisted of an insulated carrier which could be brought close to an electrified body to which it imparted its charge giving the latter an opposite charge which increased at every rotation.

A comparatively recent machine invented in 1933, but which is important to describe at this juncture because it embodies the most significant of the electrostatic principles discussed heretofore, is used world-wide in physics research laboratories, particularly for generating the very high voltages necessary to accelerate particles to achieve high energies in atomic research. The latest machine at Hartwell produces atomic particles with energies of 12 MeV. The Van der Graaf machine is described next.

THE VAN DER GRAAF GENERATOR
HISTORICAL INTRODUCTION

About 1930, an American scientist, Robert Van der Graaf, read an account of Lord Kelvin's work on electrostatic machines and was inspired to design a modern electrostatic generator. The prototype was about a foot high and made his hair stand on end when he gripped the terminal of the instrument, showing that a very high voltage had been reached:

PLATE 6 A Van der Graaf Generator. *Source Wikipedia*

The principle of the Van der Graaf generator is illustrated below:

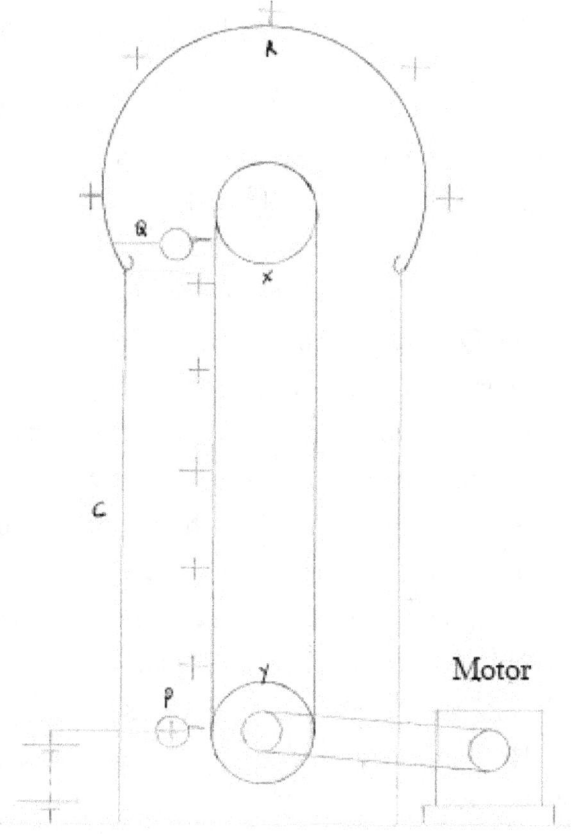

FIG 2: THE PRINCIPLE OF THE VAN DER GRAPH GENERATOR

A silk insulating belt S revolves continuously round rollers X, Y. The upper roller X is inside a large metal hemispherical shell R at the top of an insulating column C. A pointed conductor Q is placed opposite the belt near X and joined to R. P is joined to the positive terminal of a high voltage supply B, for example 10 kV.

As in the case of a lightning conductor, the pointed conductor P begins to spray its positive charge on to the belt in front of it. The charge is then carried to the top of the belt. When it comes opposite to the pointed conductor Q, a negative charge is induced on Q and a positive charge on R. The negative charge is then sprayed from Q on to the belt, neutralising the positive charge on the belt, leaving it uncharged as it passes round X. Where the belt comes round to P again the action is repeated. The net result is that R gains an increasing positive charge and hence the machine an increasingly high voltage. As noted above, van de Graaf generators can produce several million volts and as at Hartwell, are used to accelerate atomic particles in laboratories at Cambridge, Birmingham and London.

Because it embodies many of the important electrostatic principles discussed earlier, and is popular in demonstrating electrostatic experiments in education, the author decided to make a Van der Graaf machine to gain insight into problems of design and building which early electricians might have encountered in constructing electrostatic generating machines. The details are in Appendix A. Sparks up to a length of several cm in air were obtained from the replica van der Graaf generator, corresponding to a voltage of at least 40,000 volt. The experimental procedure used to determine this voltage is also described in Appendix A.

THE DISCOVERY OF ELECTRICAL CONDUCTION

Just a few decades after Gilbert, von Guericke increased knowledge of electrostatics by performing new experiments with Gilbert's 'electrics'. He found that electrics rubbed with a dry cloth of catskin became possessed of the same 'electric virtue' if they were of the same material, and tended to push each other apart. von Guericke had made a discovery which completely eluded Gilbert. An equally important discovery was that of electrical conduction. Up to that time, it had been thought that 'electric virtue' on rubbed electrics was truly static and not communicable to other objects. von Guericke showed with his electrostatic machine that it was possible to transfer the electric virtue to many materials on touching the globe of his electrostatic machine, or even through the air when a flash of light accompanied by a cracking sound were observed.

THE LEYDEN JAR

PLATE 7. THE LEYDEN JAR. *Source: Science Museum*

Its purpose being to store electric charge, in its original form (a later variant is shown above), the Leyden jar consisted of a bottle of water with a nail stuck through its cork. Its discovery was somewhat fortuitous. The gradual loss of electrification of a charged body had been attributed to evaporation. Based on this rationale arose the idea of putting the charged body into a container to reduce this 'evaporation'. A bottle was used containing water which was to be charged. A nail through its cork was a convenient method of establishing electrical contact without opening the bottle. The behaviour of the charged jar was first observed by von Kleist, Dean of the cathedral in Kammin, Pomerania, which he writes about in a letter of 1745. Holding the nail to an electric machine with one hand, he then withdrew the jar and touched the nail with his other hand, whereupon:

I receive a shock which stuns my arms and shoulder

Shocks from electrical machines were common and gave only mild pain, but this device was seemingly capable of 'condensing' large amounts of electricity, giving a powerful electric shock. The Leyden jar was later used to show that electricity could be transmitted over long distances and to demonstrate the heating, chemical and magnetic effects of electricity. Peter van Musschenbroek, Professor at the University of Leyden made another condenser out of a thin glass bowl. The shock from it struck his arms, shoulders and breast, so that he lost his breath and was two days recovering from the effects. He wrote to a friend that he would not take another for the Kingdom of France.

In 1745 Watson repeated van Musschenbroek's experiment and attempted an explanation for the Leyden jar's action. In his paper to the Royal Society in 1746 he writes:

> *The experiment is that in a vial of water is suspended a gun barrel by a wire let down a few inches into the water through a cork; and this gun barrel, suspended in silk lines, is suspended so near an excited glass globe that some metallic fringes inserted into the gun barrel touch the globe in motion. Under these circumstances a man grasps the vial with one hand and touches the gun barrel with the other. Upon which he receives a violent shock through both his arms, especially at his elbows and wrists, and across his breast. The experiment succeeds best: 1. when the air is dry; 2. when the outside of the vial is perfectly dry; 3. when the vial containing the water is of the*

thinnest glass; 4. in proportion to the number of points of non-electric contact. Thus if you hold the vial only with your thumb and finger, the snap is small; larger when you apply another finger, and increases in proportion to the grasp of your own hand; and 5. when the water in the vial is heated, which being warmer than the circumambient air, may not occasion the condensing and floating vapour therein upon the surface of the glass.

Explanations of the Effects of the Leyden Jar

Watson showed that the conductor could be a sword, a solid piece of metal or six men standing on wax cakes and touching each other. He showed that iron filings could replace water in the jar, but the effect is lessened. He also tried mercury instead of water and the strength of the shocks he received from it were the same as those from the water-filled jar. Watson explained that the effect of the Leyden jar 'arises from electrifying the non-electric water included in the originally electric glass. Hence, anything less originally-electric, for example condensation on the outside, 'defeats the experiment by preventing the requisite accumulation of electrical power'. He found that if he charged the jar with a generator and removed it without touching the connecting wire:

> *The electrified water will contain its force for many hours, may be conveyed for several miles, and afterwards exert its force upon touching the wire.*

Watson, then, did not realise that it was not the water that was electrified but the glass. Dr. Bevis improved the area of contact of the 'outer conductor' as Watson suggested by wrapping the outside with a very thin layer of lead. The new device gave a more powerful shock than the original.

BENJAMIN FRANKLIN'S EXPERIMENTS: The Discovery of Dielectric Polarization

It was Franklin who was to develop a much more convincing theory of electricity and explanation for the Leyden jar's power. In his experiments with 'M. Musschenbroek's wonderful bottle', he discovered that when a non-electric is electrified, the 'electric fire' accumulates on the surface and forms an electrical atmosphere around it. But in the Leyden jar, the glass confines the electric fire save for the surface of the non-electric (water, lead shot, etc.) and so the electric fire must be crowded into the substance of the non-electric.

This first attempt by Franklin to explain the action of the Leyden jar was made in his letter of July 1747 to a fellow physicist Collinson. In a later communication to Collinson of April 1749 entitled 'Further Experiments and Observations in Electricity', Franklin tells of his experiments with a 'dissectible condenser' by which he established that

> *...the whole force of the bottle, and power of giving a shock is in the glass itself; the non-electric in contact with the two surfaces, serving only to give and receive* [electric fire] *to and from the several parts of the glass; that is, to give* [it] *on one side, and take* [it] *away on the other*

In other words, Franklin had thus discovered the phenomenon of the polarisation of electric charge in dielectrics. He went on to make parallel plate capacitors and 'electrical batteries' out of them.

In his further 'Observations and Experiments', Franklin mentions Kinnersby's discovery using Leyden jars 'the wrong way round' by bringing the outside conductor of the jar into contact with the electric machine instead of the central conductor. Charging two jars by the usual and new method respectively, Franklin discovered that the outer and inner conductors became charged with opposite kinds of 'electric fire': 'The hook [central conductor] was disposed to give the fire, the coating [outer conductor] to receive it'. With his 'dissectible condenser', by means of a cork sphere suspended by a silk thread, he showed the charges' polarity in terms of 'positive' and 'negative' as they appeared on the 'hook', 'coating' and 'vial'. He also showed that the water in the jar did not hold any charge. The fact that a Leyden jar would retain its charge for long periods he said was due to the 'electric fluid' being resisted by the particles of the air. It keeps the electric fluid confined in an atmosphere which would otherwise tend to be leak out owing to the mutual repulsion of the particles, and would actually dissipate in a vacuum.
On the variation of the Leyden jar's efficiency with

changing thickness of the glass vial, Franklin wrote,

> *'it is amazing to observe in how small a portion of glass a great electrical force may lie'.*

Franklin certainly wondered at this observation since common sense would suggest that the larger the piece of glass, the more 'electric fire' could be 'crowded into it'.

In its final form of about 1890, the Leyden jar consisted of a glass jar lined inside and outside with tin foil. A brass knob fixed on the end of a stout brass wire passing downward through the lid on top of dry, well-varnished wood communicated by means of a loose length of brass chain with the inner foil coating. The jar was charged by connecting the outer coating to earth and the brass knob to the terminal of an electrical generator. A safe means of discharging the jar with discharging tongs, consisting of a jointed brass rod provided with brass knobs and a glass handle, was introduced by Franklin.

Although Musschenbroek had received a powerful shock from his Leyden jar, it suffered from poor insulation since one conductor was water; the insulator or dielectric was thick ; poorly conducting glass was used and a moist hand providing a low contact area acted as the second electrode. The early Leyden jar was therefore inefficient and temperamental in its operation. Nevertheless, the basic form of the Leyden jar has remained unchanged in over 200 years . A replica Leyden jar was made to investigate the problems of construction involved and its electrical performance. This is described in Appendix B.

The capacitance measured was small by modern capacitors' standard- 90 pF, but it was found that the replica Leyden jar could be charged to a maximum voltage of at least 32 000 volt.

As a storage device or 'reservoir' of electricity, the Leyden jar 'condenser' was as important in the story of the electrostatic generation of electricity as the 'galvanic' electro-chemical 'battery' was in the evolution of the electromagnetic generation of electricity, as will become clear in the next chapter.

We now turn to the historical background of the first battery.

4. ELECTROCHEMICAL ELECTRICITY GENERATION

GALVANI AND 'GALVANISM'

The foregoing has been concerned with the production of electricity by friction, and although as shown in the case of the van der Graaf generator, the voltages produced may be many thousands of volts, the currents produced are tiny. The inadvertent discovery by Galvani that electricity could be produced in frogs' legs was the first example of the production of electricity by chemical means, an effect put to use by Volta to produce currents of magnitude far in excess of any produced by electrostatic generators.

Luigi Galvani, a professor at Bologna University specialising in biology between 1791 and 1792 made the discovery whilst dissecting a pair of frog's legs. An electrostatic machine was nearby and a spark jumped from the machine to a scalpel which touched the frog's legs. The frog's legs contracted violently.

Galvani's Explanation of 'Galvanism'

Galvani's misinterpretation that the passage of the spark to the scalpel without a conducting path was the result of some power in the frog's legs led him to investigate the effect using atmospheric static electricity. During a thunderstorm, he hung his frog's legs from brass hooks which were suspended from an iron trellis. Regular muscular contractions were observed. When he took the experimental set-up inside he observed the same effect.:

> *But when I transferred the animal to a closed room, had laid it on an iron plate and begun to press the hook which was in the spinal cord against the plate, behold, the same contractions, the same motions! I repeated the experiment by using other metals at other places and on other hours and days with the same result, only that the contractions were different when different metals were used, being more lively for some and more sluggish for others. At last it occurred to us to use other bodies which conduct electricity only a little or not at all, made of glass, rubber, resin, stone or wood and always dried and with these nothing similar occurred, no muscular contractions or motions could be*

> *seen. Naturally such a result excited us in no slight astonishment and caused us to think that possibly the electricity was in the animal itself.*

Alessandro Volta repeated Galvani's experiments and concluded that the frog's body played no part in the production of electricity; rather it was a chemical effect. Although the work of Volta somewhat overshadowed Galvani's discovery, batteries were known for hundreds of years afterwards as 'Galvanic' batteries. Galvanism, Galvani's investigation with different pairs of metals and his unwitting use of the frog's legs as a crude voltmeter provided other experimenters like Volta with a whole new area of research. Volta's work is considered next.

VOLTA

Volta's Explanation of Galvanism

Alessandro Volta of the University of Pavia rejected Galvani's interpretation that electricity was in the tissues of the frog:

> *It has not been ascertained whether these movements and contractions...take place*

> *because that* [electric] *fluid directs itself by its own energy or on the other hand by the simple force of the organs of the animal toward such and such a part.*
>
> *In the latter case it may be denominated a true electricity inherent in the animal, as is asserted by Galvani...On another hand the new principle of electricity, discovered by me namely the force and virtue of metals and charcoal, in exciting and expelling the electric fluid by means of a simple contact with all humid bodies...I have established by incontestable experiments in which the bodies of animals had no concern whatever.*

Volta's researches met with immediate appreciation. As early as 1791 he was elected a member of the Royal Society London. In 1801 Napoleon called him to Paris to perform his experiments before the Institute. The French awarded him a gold medal for his work. However, there was a long and drawn out controversy between him and the equally reputable biologist Galvani over their different explanations of the 'Galvanic' effect. Galvani insisted from his experiments that the seat of e.m.f. was in the frog's body, whereas Volta rejected this in favour of his 'contact theory' and the junction between dissimilar metals as the seat of e.m.f.. Towards the end of his life, Galvani finally accepted the contact theory, but still insisted on the existence of animal electricity in the nervous and muscle systems. He showed that muscle contractions could be produced without the use of metals. Later Nobili showed that when the nerve and muscle of a frog are connected through water to a sensitive galvanometer a current is produced which lasts for several hours. He even arranged frogs' legs in series to increase the current. Recognition of 'galvanism' in the early medical world led to the use of mild electric shocks to restore people from drowning by contracting the diaphragm and chest muscles to restart breathing. A whole new field of curative medicine by 'galvanic appliances' grew.

THE VOLTAIC PILE

In a letter dated 1800 to the president of the Royal Society Volta writes:

> *I provided myself with several dozens of round plates or discs of copper, of brass or better of silver an inch in diameter more or*

> *less (for example, coins) and an equal number of plates of zinc...I further provided a sufficiently large number of discs of*

cardboard, leather or some of spongy matter which can take up and retain much water or liquid in which they must be soaked if they are to succeed I place horizontally on a table as base one of the metallic plates, for example one of the silver ones, and on this first plate I place a second plate of zinc; on this second plate I lay one of the moistened discs; then another plate of silver, followed by another of zinc, on which I place again a moistened disc. I thus continue...to form from several of these steps a column as high as can hold itself up without falling.

Volta then tried the effect of his battery on the senses. He found that in the presence of moisture when parts of the body came into both electrodes he felt a prickling sensation in the skin and activation of the muscles, an effect which he avidly compared with the twitching of the muscles in frogs' legs under the action of an electric current. Volta quickly realised that he had discovered a new kind of electricity, *current electricity*, and that it differed from *electrostatic current* in that current was continuous. However, because the potentials involved were much smaller than those produced by electrostatic generators and the Leyden jar, the effects were less spectacular:

> *My battery infinitely surpasses the power of these jars in that it does not need, as they do, to be charged in advance by means of an outside source; and in that it can give the disturbance every time that it is properly touched, no matter how often.*

He went on to investigate e.m.f.s produced by other pairs of dissimilar metals. Volta showed that the contact of two dissimilar metals in air produced 'opposite kinds of electrification', one becoming positive, the other negative. To show this, he constructed his 'condensing electroscope'.. This consisted of a gold leaf electroscope (Plate 1) combined with a small condenser. A metal plate of the element to be tested formed the top of the electroscope, and on this was placed the second metal plate which was furnished with an insulating handle. The two plates were insulated by a thin layer of varnish. A high capacitance resulted from this, enabling a very feeble quantity of electricity to charge it without altering the P.D. between the plates. When the upper plate was removed, the capacitance of the lower plate fixed to the electroscope decreased dramatically, an effect compensated by an increase in potential (In modern terms, Q=CV, and since capacitance C is decreased but charge Q remains the same, V increases). Volta made a compound bar of zinc and copper and later other metals, soldered together and held it in a moist hand. Volta's 'pile' is shown below:

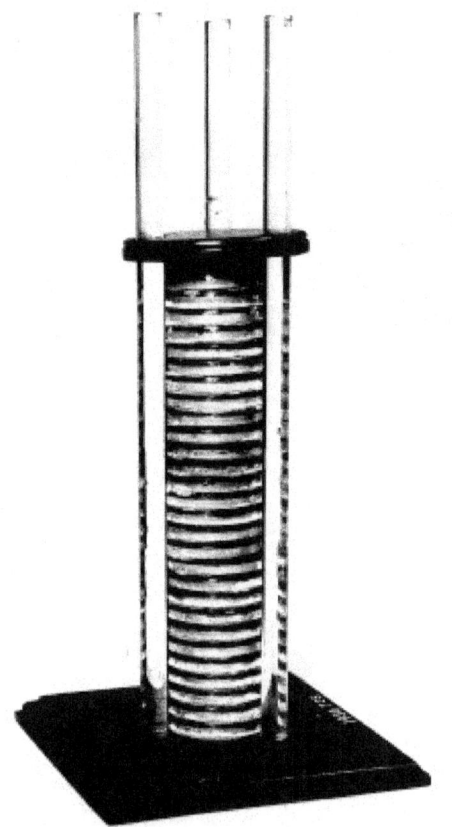

PLATE 8. VOLTA'S 'PILE'. *source: Science Museum*

One end of it he touched against the lower plate of his electroscope, whilst he placed a finger on the top plate. Opposite charges collected on the plates with the result that the leaves of the electroscope diverged, showing the presence of charge and therefore a difference of potential between the top plate of the electroscope and its case. He identified the polarity of the charge on the leaves and therefore the direction of the P.D. between the plates. Experimenting with other pairs of plates of different metals, he found that the extent of the divergence of the leaves of the electroscope (and so the P.D.) was different for different pairs of metals. Thus, while zinc and lead were respectively positive and negative to a slight degree, he found zinc and silver to be respectively positive and negative to a much greater degree. He was able to arrange the metals in a series such that each one enumerated became positively charged when placed in contact in air with the one below it in the series. According to the extent of the divergence of the electroscope leaves produced he was able to give rough approximations of the differences in their potential:

+zinc   copper
lead   silver
tin   gold
iron

The origin of electrochemical potential is described in Appendix C.

Volta's battery was the earliest generator of continuous current and in view of its importance in this context I decided to make a model of a voltaic battery to discover how much electricity it was, in principle, capable of producing and for how long, and therefore to gain some insight into the problems of using the battery in landmark experiments on electromagnetism which later ensued, particularly those of Oersted and Faraday. The construction of the replica voltaic battery is described in Appendix D. A number of problems which Volta and users of his battery must have experienced are described in detail in Appendix D. These included problems of polarisation and local action. Polarisation is the production of a film of insulating hydrogen gas bubbles in each cell in the battery during the operation which was found to seriously increase the internal resistance of the battery and lower its output after a few minutes. Local action is due to impurities in the zinc plates, which produces a black insulating coating rendering the battery useless after a few minutes, necessitating complete dismantling and cleaning of the plates. When its performance was compared with a modern Leclanche cell the voltaic pile was found to be crude, inconvenient and had a vastly inferior internal resistance and ampere-hour capacity. Volta was aware of this shortcoming and experimented with other metals for the plates.

In spite of its shortcomings, when a voltaic battery consisting of twenty pairs of zinc-copper plates are connected together about 18 volt is produced into a high resistance load such as moist skin or a Leyden jar- a voltage easily detectable by these means. And Volta had pointed the way for new experiments to try new cells with new materials.

Volta had discovered a means of generating an e.m.f. which would be called upon to supply a current for Edison's telegraph system and - with many cells connected together- as a supply of high voltage for Davy's experiments in the production of the first arc lamps, the forerunners of incandescent lamps.

THE IMPORTANCE OF VOLTA'S WORK IN LATER EXPERIMENTS

For 100 years after Galvani, batteries were spoken of as supplying 'galvanic current'. But it was Volta's explanation of Galvanism and his battery that showed the way in producing measurable continuous current, from which Oersted and Faraday discovered the *magnetic effect of a current* and *electromagnetic induction*, the pillars of modern electricity generation.

The voltaic pile was an important stepping stone between electrostatic generators and electromagnetic generators. The existence of the voltaic cell made possible the discovery of the electromagnetic effect of a current by Oersted. Without batteries experiments on electromagnetic induction would have been severely restricted and effects produced virtually impossible to detect without a continuous current of sufficient magnitude. The work of Oersted and Faraday is considered next.

5. ELECTROMAGNETIC ELECTRICITY GENERATION

JOHN CHRISTIAN OERSTED: THE DISCOVERY OF ELECTROMAGNETISM

The background to Oersted's work

William Sturgeon in 1819 had discovered that a magnet could be produced by causing an electric current to circulate around a bar of ferromagnetic material, suggesting a link between electricity and magnetism. However, it remained for Hans Christian Oersted, professor at Copenhagen University after 1806 to discover the effect of an electric current on a compass needle. Oersted obtained his Ph.D. degree for a dissertation on the philosophy of Kant, which inspired him to experiment with different natural forces to show that they were interrelated. This idea had profound consequences because it was a realisation that natural forces were not empirical or separate phenomena with independent laws governing the properties of each, but they were connected. This philosophy was later extended by Faraday to postulate a reversible relationship between magnetism and electricity.

Experiments on the Effect of a Current-Carrying Wire on a Compass Needle

The following is the account of Oersted's discovery of electromagnetism which he produced for 'Annals of philosophy', October, 1820 and which is reproduced here in full because it is fundamental to subsequent work which led to the development of electromagnetic generators :

> *The galvanic apparatus which we employed consists of twenty copper troughs, the length and height of which was 12 in.; but the breadth scarcely exceeded 2 1/2 in..Every trough is supplied with two plates of copper, so bent that they could carry a copper rod, which supports the zinc plate in the water of the next trough. The water of the troughs contained one-sixteenth of its weight of sulphuric acid, and an equal quantity of nitric acid. The portion of each zinc plate sunk in the water is a square whose sides are about 10 in. A smaller apparatus will answer provided it be strong enough to heat a metallic wire to red heat. The opposite ends of the galvanic battery were joined by a metallic wire, which for shortness sake we shall call the 'uniting conductor' or the 'uniting wire'. To the effect which takes place in this conductor and in the surrounding space, we shall give the name of 'conflict of*

electricity'.

Let the straight part of this wire be placed horizontally above the magnetic needle, properly suspended, and parallel to it. If necessary, the uniting wire is bent so that it assumes a proper position for the experiment. Things being in this state, the needle will be moved, and the end of it next the negative side of the battery will go westward. If the distance of the uniting wire does not exceed three-quarters of an inch from the needle, the declination of the needle makes an angle of about 45°. If the distance is increased, the angle decreases proportionally. The declination likewise varies with the power of the battery. The uniting wire may change its place, either toward the east or west, provided it continue parallel to the needle, without any other change of the effect than in respect to this quantity. Hence the effect cannot be ascribed to attraction; for the same pole of the magnetic needle, which approaches the uniting wire, while placed on its east side, ought to recede from it when on the west side, if these declinations depended on the attractions and repulsions. The uniting conductor may consist of several wires, or metallic ribbons, connected together. The nature of the metal does not alter the effect, but merely the quantity. Wires of platinum, gold, silver, brass, iron, ribbons of lead and tin, a mass of mercury, were employed with equal success. The conductor does not lose its effect, though interrupted by water, unless the interruption amounts to several inches in length. The effect of the uniting wire passes to the needle through glass, metals, wood, water, resin,stoneware, stone; for it is not taken away by interposing plates of glass, metal or wood. Even glass, metal or wood interposed at once, do not destroy, and indeed scarcely diminish the effect. The disc of the electrophorus, plates of porphyry, a stoneware vessel, even filled with water, were interposed with the same result. We found the effects unchanged by including the needle in a brass box filled with water. It is needless to observe that the transmission of effects through all these materials has never before been observed in electricity and galvanism. The effects, therefore, which take place in the conflict of electricity are very different from the effects of either of the electricities. If the uniting wire be placed in a horizontal plane under the magnetic needle, all the effects are the same as when it is above the needle, only they are in an opposite direction; for the pole of the magnetic needle next he negative end of the battery declines to the east. That these facts may be the more easily retained, we may use this formula- the pole above which the negative electricity enters is turned to the west, under which, to the east.

If the uniting wire is so turned in a horizontal plane as to form a gradually increasing angle with the magnetic meridian, the declination of the needle increases, if the motion of the wire is towards the place of the disturbed needle; but it diminishes if the wire moves further from that plane.

When the uniting wire be placed perpendicularly to the plane of the magnetic meridian, whether above or below it, the needle remains at rest, unless it is very near that pole; in that case the pole is elevated when the entrance is from the west side of the wire, and depressed when from the east side. When the uniting wire is placed perpendicularly opposite to the pole of the magnetic needle, and the upper extremity of the wire receives the negative electricity, the pole is moved toward the east; but when the wire is opposite to the point between the pole and the middle of the needle, the pole is moved towards the west. When the upper end of the wire receives positive electricity, the phenomena are reversed f the uniting wire is bent so as to form two legs parallel to each other, it repels or attracts the magnetic poles according to the different conditions of the case. Suppose the wire placed opposite to either pole of the needle, so that the plane of the parallel legs is perpendicular to the magnetic meridian, and let the eastern leg be united with the negative end of the battery, the western leg with the positive end: in that case the nearest pole will be repelled either to the east or west according to the position of the plane of the legs. The eastmost leg being united with the positive, and the westmost with the negative side of the battery, the nearest pole will be attracted. When the plane of the legs is placed perpendicular to the plane between the pole and the middle of the needle, the same effects recur, but reversed. A brass needle, suspended like a magnetic needle, is not moved by the effect of the uniting wire. Likewise, needles of glass and of gumlac remain unacted on.

We may now make a few observations toward explaining these phenomena. The electric conflict acts only on the magnetic properties of matter. All non-magnetic bodies appear penetrable by the electric conflict, while magnetic bodies, or rather magnetic particles, resist the passage of this conflict. Hence they can be moved by the impetus of the contending powers.

It is evident from these preceding facts that the electric conflict is not confined to the conductor, but is dispersed pretty widely in the circumadjacent space.

From the preceding facts we may likewise infer that this conflict performs circles; for without this condition it seems impossible that one part of the uniting wire , when place below the magnetic pole, should drive towards the east, and the when placed above it towards the west; for it is in the nature of a circle that the motions in opposite *parts should have an opposite direction. Besides a motion in circles joined with a progressive motion, according the length of the conductor, ought to form a conchoidal or spiral line; but this, unless I am mistaken, contributes nothing to explain the phenomena hitherto observed.*

A connection between magnetism and electricity having thus been discovered, other experimenters sought to establish the existence of a magnetic-electric effect. Ten years later, inspired by Oersted's paper to the Academie de Sciences, Faraday experimenting at the Royal Institution electricity in London was sure that if electricity could produce magnetism, the converse could be done and he was convinced that it was just a matter of time before he contrived the correct experimental conditions to produce this effect.

FARADAY

Michael Faraday was born in London in 1791. The son of a blacksmith, he was apprenticed at 14 years old to a bookbinder. He was inspired by a love of science first from reading *Encyclopaedia Britannica* and later by lectures of Sir Humphrey Davy whose laboratory assistant he eventually became. Faraday devoted his spare time to studying contemporary developments in electricity and magnetism and would perform experiments carried out by the great physicists of the day. The ability of Faraday as an experimentalist was known to the then editor of the science circular 'Annals of Philosophy' Richard Phillips, a friend of Faraday's. Phillips asked Faraday to write a series of articles for his magazine on electromagnetism, in the autumn of 1821.

FARADAY'S DISCOVERY OF THE PRINCIPLE OF THE ELECTRIC MOTOR

Faraday had read Oersted's article which claimed to have found a relationship between magnetism and electricity. As we have seen, Oersted had observed movement of a compass needle in the vicinity of a current-carrying conductor. Faraday reasoned:

According to the law of action and reaction, the wire should have an equal tendency to move around the magnetic needle. The thing to do is to make the wire movable and the magnet immovable- if such as experiment can be devised.

To test this hypothesis, Faraday fixed a long cylindrical magnet in the centre of a small cup of mercury, the bottom of which was in contact with a wire leading to a voltaic battery. The mercury was a good conductor of electricity. The magnet extended a few inches above the level of the mercury in the cup. Another wire, dipping into the mercury from above and leading to the other terminal of his battery, was necessary for a complete circuit. The problem was how to pivot the wire so that it encircled the magnet without interrupting the current through it. Faraday overcame this problem by cutting off a few inches from the top of the wire and reattached it at the same point by means of a loop. The free end of the wire rested on the surface of the mercury in the cup. In this way, the wire was movable in a circle and maintained contact with the mercury. The success of this device led him to make a similar apparatus in which the wire was fixed and the magnet hinged so that the magnet could rotate around the wire as in Oersted's experiment.

THE MAGNETIC FIELD AND MAGNETIC LINES OF FORCE

Faraday experimented with a variety of different shapes and sizes of magnets with different strengths. Experimenters as early as Gilbert had plotted 'magnetic figures' using iron filings. Faraday obtained 'fixed' magnetic fields by employing a pane of glass which had previously be gummed and dried, placing his magnet underneath, sifting iron filings through a muslin bag onto the glass. A jet of steam softened the gum which on drying stuck the iron filings fast in a fixed pattern. Using this method on many magnets, he found that stronger magnets produced a denser pattern.

QUANTIFYING MAGNETIC FIELD STRENGTH

To quantify magnetic field strength, Faraday supposed that the iron filings arranged themselves along 'magnetic lines' of force and the density of these lines was proportional to magnetic field strength. This was a concept that was to prove so valuable in understanding and quantifying the production of an e.m.f. by magnetic induction.

THE FIRST ELECTROMAGNETS

Ever since 1819 when Oersted had showed that a magnetic needle was deflected by a electric current in a wire held close to it, the great scientists of Europe and America had applied their minds to the conversion of magnetism into electricity. Sturgeon in 1825 wrapped varnished copper wire in 18 turns around a horse shoe of soft iron 1 foot long and of 1/2 inch diameter. The current was supplied by a copper-zinc cell. He found that the magnet would lift 9 pounds when the current was switched on, but the magnetic force ceased when the current was turned off.

But in order to detect the tiny currents produced by electromagnetic induction, the invention of an appropriately sensitive instrument, which became known as the 'galvanometer', was required.

THE INVENTION OF THE GALVANOMETER

The galvanometer was invented by Schweigger in 1820 immediately following Oersted's discovery. He increased the effective action of the current by carrying the wire many times around the needle. He made a rectangular or circular former of wood and wrapped many turns of wire on it. In the centre was placed a light, delicately pivoted magnetic needle. Schweigger gave the name 'multiplier' to his instrument since it multiplied the effect of a single loop of wire on the needle. However, it was realised after the introduction of Ohm's Law that as the number of turns increased, resistance to the current also increased.

The galvanometer was used by Faraday in his experiments which led to the epoch-making discovery of electromagnetic induction.

FARADAY'S DISCOVERY OF ELECTROMAGNETIC INDUCTION

We have seen that Faraday reasoned that since 'electric current produces magnetism, why should not magnetism produce electricity?' He arranged two coils of wire alongside each other. An electric current sent through a coil should induce a current in the other connected to a galvanometer, but he obtained no result. He tried inserting an electromagnet in a coil of copper wire connected to a galvanometer, but his attempts were in vain. In another attempt he prepared a number of coils of copper wire, some with many turns, some with a few. Carefully, he placed strong permanent magnets inside each of the coils connected to a galvanometer. This too failed. Some months later, he attacked anew the problem of how to turn magnetism into electricity. He made a soft ring of iron 1" thick and 62 in diameter. Around one side of it he wound many turns of insulated copper wire, the ends of which were, when ready, connected to a voltaic battery. Round the other side of the iron ring he wound another coil forming a loop around a compass needle several feet away. Faraday bent down to make a connection to the voltaic battery and happened to raise his head just in time to see the compass needle move slightly from its original position and quickly return to it. Except for that slight, momentary shifting of the needle, the experiment was considered another failure. He broke the connection to the battery and 'again the disturbance of the needle', he wrote in his laboratory notebook. At this stage Faraday was unsure of the significance of these observations although in fact now he was on the right track. He set a coil of wire close to another coil connected to the galvanometer, but obtained no results. He then decided to use bar magnets again. He wondered if the bar magnet would produced the same effect as his electromagnet had.

He placed two bar magnets flush against each other so that the opposite poles were in contact. Leaving one end of the pair touching, he prised the other ends apart and inserted a short iron rod between the poles. Around the rod he wound some insulated copper wire, joining the ends several feet away in a loop containing a compass needle. The compass needle remained stationary. Faraday grasped the north pole of the magnet and pulled it away from the iron rod. The needle moved and returned to its former position. He tried the experiment with the south pole of the magnet. Each time, when contact was made or broken, the compass needle was momentarily deflected. So Faraday had discovered how to produce electricity from magnetism- by relative motion. One had to *move* the magnet. He found he could induce a secondary current in an electrically isolated coil merely by bringing up a second coil, this one carrying a current, but it lasted for a fraction of a second. It only existed at the instant the primary current started or stopped. Instead of electromagnets, Faraday made a coil of many turns and thrust a magnet into it, again observing a deflection of his galvanometer.

In 1831 he succeeded in supplying the world with a principle, and a means of utilising this principle, to furnish a machine capable of producing a continuous electric current.

FARADAY'S DISC ELECTROMAGNETIC GENERATOR

Later in 1831 after a holiday, Faraday set to work again to produce continuous conversion of magnetism into electricity, just as von Guericke had done with electrostatic induction. He rigged up a simple loop so that it could be rotated by a handle. When the rotating loop cut the lines of magnetic force of the earth's magnetic field, he obtained an electric current from it. He set a copper disc spinning between the poles of a large permanent magnet and produced a direct current in the wires leading from this primitive magneto-electric machine (see below):

PLATE 9 . FARADAY'S DISC GENERATOR.
Source : Science Museum

Faraday's magneto-electric generator was of such historical importance that I decided to make a similar machine. The aim was to discover the order magnitude of the current and voltage Faraday would have obtained and some of the problems involved in its construction and measuring its output. The construction and performance of a replica Faraday Disc Generator is detailed in Appendix E. It produced an e.m.f. of just 1 mV.

Faraday's first reaction to the output of these machines must have been that the effect produced, especially by the rotating loop, was exceedingly small, and that devices required to measure such effects would have to be very sensitive. The next section considers the question of just how sensitive and suitable early current detecting instruments, notably the galvanometer, were.

GALVANOMETER SENSITIVITY

The Schweigger multiplier was the earliest galvanometer available to Faraday. Appendix F performs calculations which show that, equipped with such an instrument, Faraday could indeed detect currents from his disc machine which were as small as a fraction of 1 mA.

FARADAY'S LAW AND THE IMPORTANCE OF FARADAYS DISCOVERIES

Faraday's discovery of electromagnetic induction pointed the way for future experimenters. He showed that the e.m.f. induced was intensified by using more turns on his armature and using a stronger magnetic field. Finally, he discovered the principle of the transformer, also of vital importance in today's electricity supply industry. Faraday revolutionised the concept of magnetism when he quantified the intensity of the magnetic field; enhanced a magnetic field many hundreds of times using soft iron armatures; and discovered the vital principle that the *rate* of cutting lines of magnetic force produces an e.m.f. according to the law bearing his name:

$e = -d\phi/dt$, where $\phi = BA$ is the magnetic flux.

Faraday's disc generator with an output of only about 1 mA, did not produce any commercial interest: it certainly could not compare with that of the chemical cells used by Davy at the Royal Institution. But Faraday's discoveries showed how the principle of electromagnetic induction might be enhanced to produce a much greater effect. It was at this stage of the evolution of the generator that electromagnetic induction could be fully understood, building upon the concepts of magnetic field and field intensity introduced by earlier experimenters, paving the way to the generation of larger quantities of electricity. It was in the hands of later physicists to develop Faraday's ideas, beginning with Hippolyte Pixii.

MAGNETO-ELECTRIC AND EARLY ELECTROMAGNETIC GENERATORS

THE PIXII GENERATOR

French instrument maker Hippolyte Pixii's magneto-electric generator of 1832 was an early attempt to commercialise generators based on the principle of electromagnetic induction (see next page). In this machine a permanent magnet was rotated via gearing by hand, inducing a current in two coils wound around iron armatures, which greatly enhanced Faraday's electromagnetic effect by using many turns of wire instead of a single loop. It produced a much greater e.m.f. than Faraday's disc generator, even though restricted by the crude method of applying the weak magnetic field. The construction and performance of a replica Pixii generator is described in Appendix G.

The performance of the Pixii generator was a very considerable improvement on the Faraday Disc Machine. The O.C. e.m.f. obtained was 0.25 volt, 25 x that of Faraday's machine, and a current of 24 mA into a 1Ω load was obtained, almost 50 x the output of the replica Faraday Disc Machine.

PLATE 10. PIXII GENERATOR. *Source: Science Museum*

Pixii's generator was sold as little more than a toy: such devices were popular with the well-to-do at this time. But it was the forerunner of a line of generators whose output thereafter rose sharply, as more turns of wire and stronger magnetic fields were used in new, more efficient configurations. After about 1840, physicists began to realise that the electromagnetic generator might, with further development, constitute a more effective source of electrical power than batteries for Davy's arc lights. Eminent physicists then turned their attention to this problem.

OTHER EARLY ELECTROMAGNETIC GENERATORS AND THEIR SHORTCOMINGS

In the year following Pixii's generator, a machine was demonstrated to the newly formed British Association for the Advancement of Science by Joseph Saxton in which the magnetic field was still provided by a permanent magnet, but the e.m.f. was induced in a rotating armature while the magnet was fixed. This generator, as did Pixii's, had the disadvantage that *A.C.* was produced, which at first made them virtually useless as a direct replacement for batteries used by scientists to produce steady *D.C.*.

THE INVENTION OF THE COMMUTATOR

Then a device called a commutator was invented by William Sturgeon in 1834, which maintained the current in one direction. This was accomplished by terminating each coil connection on opposite sides of a metal cylinder consisting of two halves insulated from each other with mica. The two output terminals were flexibly connected to each half of the commutator. The flexible connections to the cylinder were placed so as to change position from one half to the other when the direction of the current began to change. The figure below shows commutated D.C.:

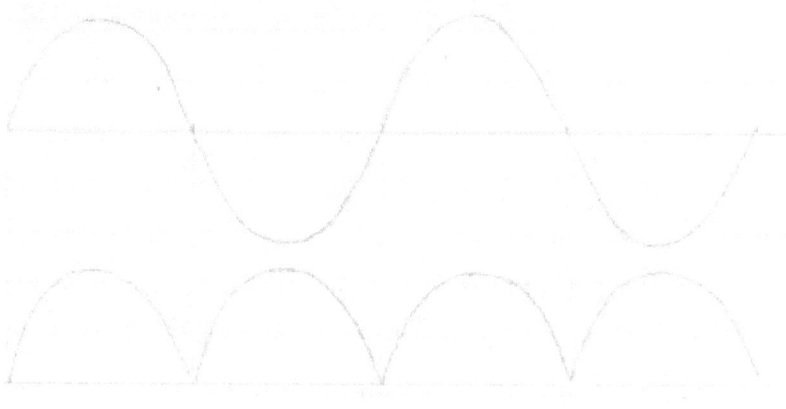

FIG. 3 COMMUTATING ALTERNATING CURRENT

The upper diagram shows A.C., lower diagram commutated A.C.

As can be seen from the lower diagram, the commutated output from a two-pole generator was far from a steady D.C.. An obvious remedy to try was to increase the number of pairs of poles. Wheatstone in 1841 used a multiplicity of poles on a rotating armature which produced a much smoother output. However, as generators of electricity early machines were still very inefficient.

SELF-EXCITED GENERATORS

An important advance was made by C.F. Varley in 1866 when he discovered the principle of self-excitation. Varley's principle was that part of the e.m.f. produced a generator could be diverted to the coil of an electromagnet to produce the magnetic field of the generator. The principle produced strong magnetic fields to the armature coils which improved output considerably: it also made this type of generator self-contained. But to produce a useful output, generators had to be large to carry large electromagnets and armature coils with many turns. It therefore became apparent that the method of revolving generators by hand was impracticable with machines above a certain size.

Varley's self-excited generator of 1866 utilised the basic ideas of construction which are used in modern generators. High density magnetic fields: self-excitation and multicoil armatures are being used today. But the armature in particular as a means of disposing windings compactly, enhancing the effect of the magnetic field and of supplying a sustained output, was to receive a great deal of attention. It is to the evolution of the armature that we turn next.

EVOLUTION OF THE ARMATURE

THE GRAMME RING

1870 marks the start of the era of great progress in the design of generators capable of supplying electricity in large enough quantities to work arc lamps with sufficient reliability over long periods. In this year, Z.T. Gramme introduced an armature which consisted of a ring comprising many soft iron wires around which coils were wound in several layers, covering the whole surface of the soft iron armature. The coils were separately terminated on metal segments fixed in an insulated drum from which the generated e.m.f. was picked up by means of brushes:

PLATE 10. EARLY GRAMME DYNAMO.
Source: Science Museum

EDDY CURRENTS AND HEAT PRODUCTION

Gramme's machine was limited because in cutting the magnetic field the soft armature had induced in it an e.m.f. parallel to the direction of the windings of the coils. Because the ohmic resistance of the armature was low, these induced currents, know as 'Eddy' currents after their discoverer, were allowed to build up and produce a strong heating effect in the armature. Because of the low surface area of the armature core, being almost entirely covered by several layers of copper windings, this heat could not dissipate effectively and overheating, damaging the insulation of the windings, and became a serious problem. An armature in the shape of a drum was introduced in 1876 by Germans Siemens and Halske on which the layering of windings was reduced by the increased surface area and its enhancement of the magnetic field. But the early armatures of Gramme and Siemens and Halske suffered from the destructive effects of centrifugal forces which tended to loosen windings, so that their speed of revolution was restricted. Also, the armatures could not keep pace with increasing demands made on their output by commercial users owing to the cooling problem. The Swiss engineer Burgin improved cooling by arranging the armature in the shape of a turbine rotor which drew air through it, improving ventilation The armature rotor blades, called spiders, seen in the figure on the next page, were staggered along the length of the armature and each spider carried a hexagonal core consisting of many strands of soft iron wire:

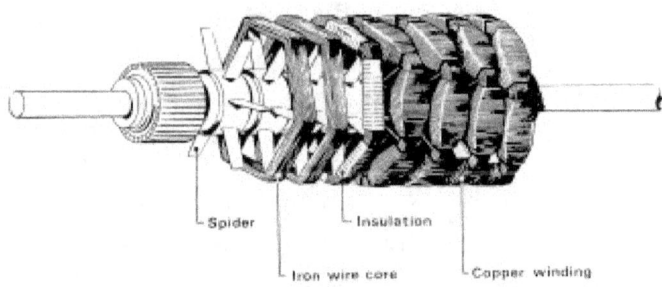

Crompton-Bürgin armature, showing its internal construction

FIG. 4 .CROMPTON-BURGIN ARMATURE: *Source: R.E.B. Crompton*

The cores were lapped with one layer of coils wound at right angles to the line of the core. Each coil on the armature was connected to the nearest coil on an adjacent spider and each spider carried six coils. The low mass of the spider cores and security of the windings meant that the armature would not be damaged by the centrifugal forces of high-speed running.

SEPARATELY EXCITED FIELD GENERATORS AND LAMINATED CORES

In the 1880's, Burgin type armatures were used in Crompton-Burgin generators designed by R.Crompton:

Crompton-Bürgin generator

FIG. 5 CROMPTON-BURGIN GENERATOR:
Source: R.E.B. Crompton

The armature revolved in a usually separately excited electromagnetic field energized by a single Gramme generator. But Crompton soon realised that the utilization of a large cross-section of iron in the armature was made difficult by the design of the Burgin armature. He then turned to the conventional Gramme and drum-type armatures which he improved the effectiveness of by the use of better materials and other refinements. A *laminated* core and conductors embedded in the armature were introduced, which not only increased surface area and effective field strength, but by separating the thin iron sheets of the core by coats of varnish the ohmic resistance of the core was increased considerably, greatly reducing the circulation of Eddy currents within it and therefore the heat generated. [see Trigger (2013) for more details]

Prior to the 1880's machines were fitted with commutators to convert their output from A.C. to D.C., for use with arc lights and other applications. But increasingly, engineers began to realise the advantages of A.C. over D.C., particularly as electricity had to be conveyed over increasing distances with minimum losses as supply became more widespread . The long debate which arose of over the merits and demerits of A.C. and D.C. generation and transmission became known as the *Battle of the Systems* and is discussed further later.

SELF-EXCITED VS SEPARATELY EXCITED FIELD GENERATORS

Prior to 1882, there existed two methods of energizing the magnetic field windings of generators. The first was used in the Crompton-Burgin machines in which the field was energized by a separate Gramme generator, as we have seen: it was separately excited. The second method, introduced by Varley, passed all of the armature current ('series wound') through the field magnet coils or part of the armature current was diverted through the field magnet coils ('shunt wound'):

PLATE 12. VARLEY'S SELF-EXCITED GENERATOR. *Science Museum*

Both types of self-excitation had their merits, and so in 1882 Crompton and Kapp introduced 'compound' winding in which the advantages of both were incorporated to give a steady output.

In Appendix H, the two methods of excitation are technically appraised.

In the next chapter, we move on to consider electric lighting which drove the burgeoning need for electricity supply.

6. ELECTRIC LIGHTING
THE NEED FOR SUPPLY

EARLY ARC-LIGHTING BY BATTERY

In 1802, using 2 000 voltaic cells at the Royal Institution in London, Sir Humphrey Davy discovered that when a spark is struck between two carbon rods a brilliant white light was emitted: he had discovered the principle of the arc lamp. Interest in Davy's invention was held back until years later, when an economical source to power it became available. However, after 1831 following Faraday's discovery of electromagnetic induction, engineers saw the potential of Davy's discovery as a source of light to replace the then ubiquitous smoky and dim gas lighting. From then on work was begun by ambitious men like Siemens, Gramme, Crompton and Ferranti to produce machines capable of generating enough electric current to supply arc lamps for lighting the premises of commercial, public, and later, private consumers.

LIGHTING OF LIGHTHOUSES BY ELECTRIC GENERATORS

The advent of the supply of mechanically generated electricity to energise arc lamps was initiated by the English engineer Frederick Holmes. It was suggested that an important use of arc light would be to use it for lighthouse illumination to make coastal areas safer for seamen, and Holmes used a generator designed by Meritans. The French engineer Baron A. de Meritans was the first to demonstrate a generator driven by a *steam engine* before the Brethren of Trinity House in 1857. His demonstration of the $1\frac{1}{2}$ kW steam-powered generator was a culmination of work on a project to construct a suitable generator to supply arc lamps for lighthouses. The success of this demonstration led to its immediate adoption in the South Ireland lighthouse. Four years later, Dungeness lighthouse was similarly equipped.

OVERCOMING THE SHORTCOMINGS OF ARC LAMPS

Early arc lamps contained carbon of insufficient purity and hardness for the electrodes with the result that they burned away quickly, so that the gap increased and the light output was reduced. The period between adjustments was elongated by using better quality materials, but what was needed was a mechanical device to automatically regulate the air gap between the electrodes. O.K. Staite since 1846 had striven to invent such a device, and succeeded in producing an arc lamp in which the gap was regulated by the expansion of a copper rod by the heat of the arc, which narrowed the gap between the electrodes as they burnt away. During the 1870's when Gramme's ring dynamo came on the scene as a satisfactory source of electrical power for a series of arc lamps, greater interest was taken in the arc lamp for general lighting and wider application followed, beginning with public buildings/works and large premises. A new and improved design was introduced by P. Jablochkoff in 1876, a Russian engineer. His lamps consisted of vertical, parallel carbon electrodes instead of being placed end to end. There were no special problems of regulation of the gap since the lamps were run on A.C. which meant that the electrodes burned away at the same rate.

EARLY MUNICIPAL ARC LIGHTING

In 1877 eighty such lamps were installed in a store in Paris and in the West India Dock, Billingsgate

Market and Holborn Viaduct, and part of the Thames Embankment was lit with Jablochkoff lamps.

Further Improvements to the Arc Lamp

Crompton, a pioneer in electric lighting, became interested in electrical engineering in connection with the Stanton Iron Works in Derbyshire. This connection was a result of its ownership by relatives. In 1878 he designed a foundry for cast iron pipes, but in order to be economical the foundry had to be worked day and night.

Dissatisfied with existing arc lamps, Crompton designed his own. These were equipped with an electromagnet mechanisms which regulated electrode air gap automatically:

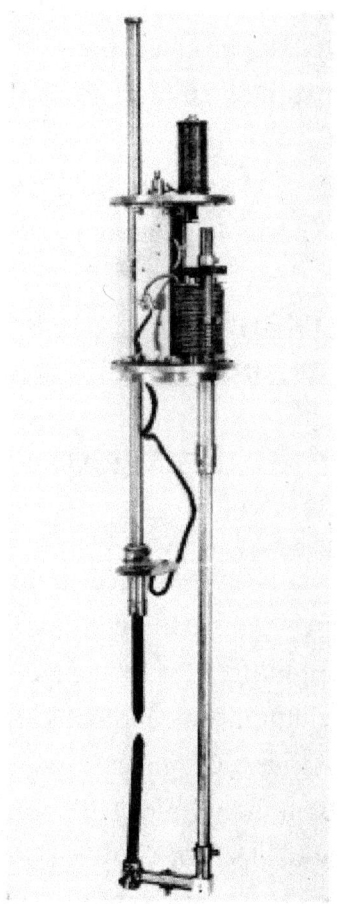

The earliest surviving Crompton lamp

FIG 6 . CROMPTON ARC LAMP. Source: R.E.B. Crompton

They worked quite well, but when Crompton started to produce them commercially 1880 they sold at the then high price of £15.

At first, electric lighting was carried out by small, local portable generators, as described below.

CROMPTON'S ELECTRIC LIGHTING SYSTEMS: LOCALISED SUPPLY

PORTABLE GENERATING SETS FOR ARC LIGHTING

As we have seen, electric lighting of 4 lighthouses by arc lamps had been in existence since 1858, but the lighting of public places commenced with the installation of Jablochkoff lamps powered by Siemens A.C. generators in 1878 at the Victoria Embankment and Holborn Viaduct, although as early as 1875 a Gramme generator was supplying power to arc lamps at the Gare du Nord in Paris.

But the arc lamp eventually gave way to the incandescent filament lamp which provided a less dazzling glow, had a longer life, was cheaper and consumed much less electricity.

Following the introduction of the Swan lamp in England and improvements in generators, electric lighting companies began to spring into existence, mainly supplying street lighting and lighting for commercial users such as shops and small factories, selling surplus electricity for the private homes of the nobility. By the late 1880's small electric lighting companies were dotted all over the urban area of London.

CROMPTON'S LIGHTING OF IRON WORKS

As we have seen, to light his iron works, Crompton used his own arc lamps energised by imported Gramme generators, each lamp being supplied by one generator. Crompton sustained such interest in studying electrical engineering and lighting his ironworks that he decided to go into production of Crompton arc lamps and import Gramme generators from France, which he did in premises at Anchor Ironworks at Chelmsford. In a few years owing to great success, Crompton took over the ironworks at Chelmsford and started production of machines equipped with Crompton-Burgin armatures.

CROMPTON'S LIGHTING OF PUBLIC EVENTS, BUILDINGS AND RAIL STATIONS

To advertise his electric lighting system, Crompton and his assistants travelled in the Home Counties hiring out portable generating sets:

Portable generating set to work at night and for fêtes and other public entertainments.

FIG. 7. PORTABLE GENERATOR. *Source: R.E.B. Crompton*

These were used at public entertainments and sites where it was necessary for work to continue throughout the night. The first generating set produced in 1879 consisted of a horse drawn steam engine coupled to the small early Gramme generator via a pulley belt. One of the first public events to be lit by Crompton was the Henley

Regatta of 1879 which attracted considerable public interest. Following this, two Crompton arc lamps suspended 30 ft above ground lit the fete of the Butchers Provident Institution at Waltham Green and four arc lamps supplied by four Gramme generators ware installed in the grounds of Alexandra Palace:

The Crompton Light on the Lakes at Alexandra Palace

FIG. 8 CROMPTON'S ARC-LIGHTING AT ALEXANDRA PALACE. Source: R.E.B. Crompton

During Christmas 1879, Crompton lit his own house in Porchester Gardens in London using one of his portable generating sets parked in the mews near to his house.

After two successful years of supplying electric lighting for outdoor activities the booklet 'The electric light for Industrial uses' was published, in which Crompton exhibited and explained his system in detail. But orders coming in by 1881 required more ambitious installations. The British Electric Company used 6 Crompton arc lights of 6 000 c.p. each supplied by a Gramme generator for lighting St. Enoch's in Glasgow. These replaced the 464 gas jets, giving a better light with similar running costs.

Crompton and Company Limited was contracted in 1881 to illuminate the sorting room at the Central Post Office in Glasgow. This, the first industrial indoor installation by Crompton was an experiment to test the running of this lamps in a confined space. The sorting room measuring 114' x 54' x 25' was lit by two Crompton arc lamps supplied by portable generating sets. These replaced 180 gas jets and provided better light, a cleaner atmosphere and lower temperature in the room. Crompton's distribution circuits of this period consisted of bare copper conductors supported on insulators, and all his installations using early Gramme generators comprised two wire circuits between each lamp and generator.

In 1883 Crompton lit Kings Cross station with his arc lights. This was the first large installation in which Crompton-Burgin machines were used with more than one lamp per generator. 12 arc lamps of 4 000 C.P. were suspended in two lines of six above the platform and supplied in threes to each generator. A further two lamps were installed in the station forecourt and supplied by a fifth generator. All five machines were driven by a single steam engine. The new law courts erected in the Strand in 1882 were also lit by Crompton, although Swan incandescent filament lamps were used instead of arc lamps. Arc lamps were to be almost universally replaced by incandescent lamps, but as the work of Edison and others described below shows, the development of long-lasting lamps based on this principle was beset with problems of filaments burning up quickly at white heat.

THE INCANDESCENT LAMP

Staite was the first to demonstrate an incandescent filament lamp in 1847 using a filament consisting of an alloy of platinum and iridium enclosed in a glass envelope evacuated of air. However, the filament lasted only a short time because there was no effective means of exhausting all the air from the bulb. Another problem was to find a suitable material for the filament which would stand up to white-hot temperatures for reasonably long periods. A significant step forward came with the invention of the mercury pump in 1865. But the problem was a source of many hours of experimentation for Thomas Alva Edison in America in the late 1870's.

EDISON'S AND SWAN'S INCANDESCENT LAMPS

Edison had read about the use of Jablochkoff 'candles' to light the square around the Theatre Francais in France. The square was brilliantly lit by 64 arc lamps and made the area at night for hundreds of yards around it as light as day. But the light fluctuated and changed colour. Edison had also heard about an incandescent filament lamp made by Joseph Swan in England in 1860. This consisted of a conductor formed by pieces of paper or card packed with charcoal powder in a crucible, baked in a pottery kiln and then placed in a bottle with a wide neck from which air had been partially evacuated. Swan's lamp was very dim, and the filament was excessively delicate. After the introduction of the mercury pump, a better vacuum was obtained but the filament still tended to disintegrate.

At the time of hearing about the efforts of Swan, Edison was preoccupied with his work on the phonograph. In July 1877, Edison put this aside and tried a few experiments with incandescent-filament lamps. He first tried an inch long strip of charred paper with its ends fastened to a battery. On heating up, it disintegrated. He then burned it in an exhausted vessel and this time he was able to keep it incandescent 8 minutes before it disintegrated. He tried different materials such as wood, broom corn, and other fibres, coating them with a mixture of lampblack and tar and rolling them into the form of a needle and carbonising them in a furnace. Work on his phonograph and transmitter prevented him from giving the lamp his full attention until April 1879. In his laboratory at Menlo Park he tried various filaments of rare metals coated with oxide. He succeeded in producing a lamp of high resistance, but it was prohibitively expensive. It consisted of a filament of a compressed spool of zirconium wound with a thin platinum wire and insulated with oxide. To improve the longevity of the filament, Edison tried it in a vacuum produced by a mechanical air pump. A partial vacuum of 1 mm pressure of mercury was produced, but he noticed that the pressure gauge fluctuated violently at this pressure. Edison concluded that the gases imprisoned in the metals of the filament were being exhausted. This discovery greatly increased the efficiency of his lamp. Previously only a current capable of producing a light of 5 or 6 candle power could be used before the filament burnt out, now about 25 candle power was obtainable. Edison was convinced that lamp life would be increased if he could produce a better vacuum. A new larger pump made by Sprengel was obtained. For experimental purposes, Edison employed a glass blower to blow glass bulbs with spools to carry a filament introduced in them. His zirconium-platinum filament saddle on a spool of clay in a glass envelope evacuated by the new mercury pump produced a brilliant white light.

But Edison was not satisfied with his lamp, because it was still far too expensive for commercial use or as a replacement for modern arc lamps. He had to try cheaper, more mundane materials. He tried a length of cotton smoked in a furnace for an hour. This produced a brilliant light but could not take full battery current. Under the microscope, the filament showed a very hard crystalline structure due to carbonisation. The thread was discarded and splinters of wood, straw, paper, silk fish line, vulcanized rubber and human hair were carbonized in an attempt to obtain a longer lasting filament. Difficulties arose in carbonising the materials because of the different rates of expansion of the material and the former on which it was wound, which resulted in breakage in the furnace. Those which were taken in tact from the furnace still had to be mounted in the bulb, which required considerable deftness and patience and the number of breakages was frustratingly high. Also, breakages occurred when the glass blower attempted to seal off the glass bulb. Under these conditions, just two or three experimental lamps were completed in a week.

But at last in October 1879 Edison made a lamp which produced a bright light for over 40 hours. In making this lamp, Edison decided to try again a filament made from cotton thread. He mounted the filament in a bulb and started to pump the air out. When the pump began to click loudly it was a sign that the evacuation was nearing completion. To drive out the gases from the filament and expel the remaining gases in the bulb, Edison heated it with a spirit lamp. With this filament lamp, he applied current in gradual increments to ensure that all gas was expelled- and sure enough large air bubbles appeared in the pump tube. He continued to apply battery current to the carbon filament from time to time, increasing its intensity and duration until the highest vacuum feasible was obtained. The glass blower then sealed off the bulb.

THE FIRST VIABLE CARBON FILAMENT LAMP

Watched by shifts of Edison's workers around the clock, the lamp was lit and was to continue for over 40 hours until Edison increased the voltage until in a brilliant cascade of the light, the lamp went out. The filament was immediately scrutinized under a microscope. In a short space of time, Edison improved the operating time of his lamps to 100, then 150 and 170 hours with a change of filament material from cotton to cardboard. Edison's carbon filament lamp was announced in the New York Herald, amidst a storm of inconfidence in his work from other scientists working to produce a filament lamp, on December 21st, 1879. People came from miles around to see selected homes along Christie St. where he had his laboratory lit by lamps supplied by his dynamo electric machine devised with his associate Upton.

Convinced by his calculations that electric lights could be made cheaper to run than gas lights, after further developing his lamp Edison went into production in an old railway hut. Initially he sold his lamps for 40^c which cost him $1.25 to make. His next enterprise was to convince his bank financiers, by building a small demonstration electric lighting system in Menlo Park with his dynamos and own distribution system, that he

> *...could devise the means of establishing electric lights on a commercial basis; to distribute the current from a central station and measure it, as gas is now measured, and to bring cost down to the point where the enormous moneyed influence of gas can be successfully contended.*

SWAN'S IMPROVEMENTS

Meanwhile in Newcastle-Upon Tyne, Swan was also experimenting with incandescent lamps; he had been trying new materials in new configurations in the search for a long-lasting filament since 1848, when at last he produced a carbon filament lamp with reasonable longevity in 1878. Unfortunately, Swan did not patent his inventions as did Edison, and found his way blocked. However, in 1880, he patented a method of eliminating the last traces of the gasses from filaments by the 'flashing' process. This increased the life of the lamp and reduced blackening of the bulb with age due evaporation of the filament. In 1883 Swan produced an improved filament by extruding dissolved cellulose into a thread and carburizing in a furnace.

Edison and Swan realised that their interests would be best served by cooperation and the Edison and Swan United Electric Light Company Limited was formed in 1883. The application of incandescent filament lamps was slow to spread at first until their use in famous buildings proved their worth. In 1881 the House of Commons was lit with the new lamps, and the Savoy Theatre was lit by 1 000 lamps in the same area of London. Following the lighting of the British Museum, the Royal Academy, ships and trains began to receive application. From the early 1880's, the lamps were increasingly applied in private houses.

LATER DEVELOPMENTS

The development of the incandescent filament lamp did not stop there. Inventors sought a means of making a satisfactory metallic filament. Although filaments consisting of osmium with a high melting point were introduced by von Welsbach in 1898, scientists were convinced that the best prospect was tungsten, and tungsten filament lamps were in general use by 1911. Even by 1900, electric lighting by filament lamps was a feature of London life.

LIGHTING DOMESTIC PREMISES WITH SWAN LAMPS

In the 1880's, the Swan lamp was a comparatively recent introduction and before the public became convinced of the merits of the Swan lamp compared with the arc lamp, industrial installations, particularly outdoor applications, were provided with brighter, easy to install and tried and tested arc lamps.

But for small, internal applications especially private homes, the Swan lamp was to be preferred. An early private subscriber whose house was equipped with Crompton electric lighting wrote to the *Times* about the advantages of Swan lights over gas lighting:

> *The light is as easy to manage as gas, while the softness, the purity and the agreeableness are such that a return to any other method of illumination would be out of the question. The pictures, books and decorations have no chance of injury; the ceilings and walls remain unsoiled, while the difference in health felt after sitting for an evening in a room electrically illuminated, and another lighted by gas must be experienced before it can be appreciated.*

In this same letter, this gentleman gives an account of the expenses incurred in installing the lighting system and the running costs. The lighting of his house by Swan lamps of 20 C.P. each required 220 lamps. His home in Brentwood could not be supplied by existing street lighting generators because there was none near, so that four Crompton-Burgin generators had to be installed at his premises. The installation cost was a phenomenal;- £1 500- and the annual running cost £250.

CENTRALIZED SUPPLY

Up to 1885 Crompton used low voltage to supply lamps direct from his generators. Swan lamps, for instance, ran on a pressure of 60 volt. The disjointed state of electricity supply at this time was marked by the existence of numerous small electric lighting companies using their own generators supplying mainly street lighting, although surplus electricity was sold where possible to commercial interests. Not surprisingly, electric lighting companies were thickest on the ground where the need for supply was greatest, and supply was only remotely economical within these areas. Crompton and Ferranti (whose work is discussed later) were aware of the need to provide central generating stations and a distribution network, but the government legislation required to give such ventures freedom of the way was a long time in coming. Meanwhile, those enterprises which were receiving electricity supply were experiencing very high costs, poor voltage regulation and frequently interrupted service.

Crompton was not the only engineer who had ambitious plans for the large scale lighting of London. Edison in New York had been experimenting with methods of distribution of electricity , to which we now turn.

EDISON'S LIGHTING SYSTEMS
DISTRIBUTION OF CURRENT

After developing his carbon filament lamp, Edison turned his attention to the distribution of current. His lamps had been demonstrated at his laboratory in Menlo Park and lamps were installed along Christie Street and in selected homes along it. Edison and Upton had designed a very efficient dynamo to power Edison's lamps. However, his existing methods of distribution caused too great a drop in pressure in certain parts of the network, causing the higher resistance in furthest away paths to gave rise to dim lights. The amount of copper used was also prohibitively expensive.

'MAINS' DISTRIBUTION

Edison experimented with new ways of arranging his conductors and came up with a distibution scheme in which conductors were to be laid along both sides of the street and connected at either end to junction boxes called 'mains'. These 'feeder' wires were connected to the central source carrying current from the generator. Individual premises along the street would be supplied from the mains. This system, which ran at the same voltage as the lamps, 100 volt, used 7/8ths of the copper used in the previous system and the fall in voltage under load was less in the feeder circuits which supplied current at the average voltage of the lamps.

THE CASE FOR REPLACEMENT OF GAS LAMPS WITH ELECTRIC LAMPS

To achieve the support of his financiers, Edison had to demonstrate the advantages of electric lighting over gas, the economy of the system and that there was sufficient demand for electric lighting to justify the construction of a central generating station. Edison employed men to go into New York City to record how many house holders were willing to have their gas replaced by electricity, and how much gas each pendant consumed. Consumption of the gas street lighting was also determined. Edison and his associates calculated how much horse power would be required to provide adequate lighting in each house, and after studying detailed maps of the areas to determine what routes his distribution cables would take, he calculated the amount of copper required. Edison's system was demonstrated to his financers and was a great success. Immediately, an application was forwarded to the City Hall in NewYork to obtain permission to dig up the streets in a district in the lower part of the city for the laying of underground cables.

EDISON'S LIGHTING OF NEW YORK CITY

After two years in 1881, permission was finally granted, but meanwhile Edison had been working on a dynamo which was to be directly coupled to a steam engine made by C.Porter of New York instead of using an inefficient belt drive. The dynamo and steam engine ran at 750 r.p.m. and weighed over eight tons. Edison's next step was to look around for a site for his generating station and found it in a building at 255-7 Pearl Street. Edison moved his premises from Menlo Park to Goerk

Street where he began manufacture of more dynamos, this time using the larger steam engine designed by G. Sims. To undertake the lighting of New York City the 'Edison Electric Light Company' had to reform as the 'Edison illuminating Company of New York'. As as soon as the project was under way, contracts for electric lighting came in from all over the proposed area. To cope with the demand, a new lamp factory was set up in East Newark, New York and 'runners' began wiring houses.

PEARL STREET

In July 1881, the laying of the street mains began in New York City. It was proposed to illuminate the district of Manhattan over one square mile by putting his electric lights in commercial establishments and in 2 500 houses. The mains were to be run in iron pipes under the city and work commenced at the rate of 1 000 feet of conductors per day. Work continued through the night under lights fed by small dynamos in the Pearl Street building. The 14 miles of mains were laid parallel to the curb in trenches in accordance with his feeder and mains system used at Menlo

Park. Edison's crews wired offices and dwellings, connecting them with the mains in the streets and supplied lamps and fittings and meters for measuring consumption. The entire district of Manhattan would ultimately be lit by six large bipolar dynamos directly coupled to Sims steam engines. In early September 1882, the system was ready for its first test and the main switch connecting one generator to the mains was pulled by the electrician at Pearl Street station. At this stage, only about 1/3 of the district could be lit owing to controversy between the Board of Underwriters and the Edison Illuminating Company of New York over the degree to which wiring of houses presented a safety hazard.

Also, a satisfactory means of connecting generators together had not yet been found, so that only the capacity of 1 200 lamps for one machine could be lit. At the end of a day of encouraging operation, Edison remarked:

> ...*it has been entirely successful. We have only one engine running now. It supplies eight hundred globes with light. We have six engines, and all will be in successful operation before the end of the winter.*

Before Pearl Street, except for a few spots illuminated by Jablochkoff and Brush arc lamps, New York was dimly lit by scattered gas lamps. Due principally to the conviction and endurance of one man and his work, the sceptical authorities and the public were convinced of the necessity of centralised electricity supply.

EDISON'S BIPOLAR DYNAMO

Edison was dissatisfied with the generators in existence in 1880. They were less than 40% efficient for two reasons. The resistance of the armature was made closely equal to the resistance of the external circuit which resulted in a large amount of current being used up to overcome the resistance of the armature coils. Edison recognised this inadequacy. 'It is ridiculous', he said, 'Do you think I want to lose 50% of the power I produce?' He also recognised that generators of the day tended to use too many coils on their armatures, with the result that much of the field was not properly utilized owing to magnetic saturation of the iron. He said,

> *When these fellows use too many windings of wire in a generator, and too much current for*

a small cross-section of iron, they throttle the lines of force out of the right path. I believe you can saturate a piece of iron so you can't fit another line into it.

So Edison went back to work on producing a generator in which the armature windings had as low a resistance as possible and having a large cross-section of iron to make maximum possible use the field produced by two large electromagnets:

PLATE 13 . EDISON BIPOLAR DYNAMO
Science Museum

This machine was the forerunner of his 'Jumbo' dynamos used at the Pearl Street generating station. Three well-known, independent scientists were invited to his laboratory in Menlo Park to measure the efficiency of his new dynamo. This was calculated by immersing lamps in water and determining how much heat was given out in a measured time, while the rate of working of the steam engine was measured. An almost incredible efficiency of 98% was obtained. In a later test in which Edison's lamps were coupled to his generator, the commercial viability of his electric lighting was proved when it was found that sixty of these lamps could be supplied by each (746 watt) generator. Edison Company in London set up a generating station at their premises in Holborn Viaduct to supply street lights from Newgate Street to Holborn Circus. But this system was quite small compared with the load area envisaged for the Pearl Street generating station in New York City which by the autumn of 1881 was well under way. Edison directly coupled his D.C. bipolar dynamo to a steam engine capable of supplying over 800 of his incandescent filament lamps.

S.Z. de FERRANTI
FERRANTI AND THOMPSON'S IMPROVEMENTS IN GENERATOR DESIGN

In 1882, when a young man of eighteen, Ferranti completed a design for an alternator which incorporated a revolutionary armature. An acquaintance of Ferranti, Sir William Thompson, had produced a similar dynamo at about the same time and the two men decided to co-operate to produce a machine which incorporated a field magnet of Thompson's design and an armature of Ferranti's. A company, Ferranti, Thompson and Ince Limited, was formed to manufacture the new Ferranti-Thompson Dynamo.

The introduction of the machine caused quite a stir as it was claimed that it could satisfactorily light 5 x as many incandescent lamps as any other machine of similar size. The company with sole rights of supply, the Hammond Electric Light and Power Supply Company, gave a demonstration to engineers and the press at their premises near Canon Street, London.

FERRANTI'S 'ZIG-ZAG' ARMATURE

The Ferranti-Thompson dynamo was fitted with Ferranti's zig-zag armature which was constructed of a strip of copper 120ft long and 1/2" wide but contained no iron:

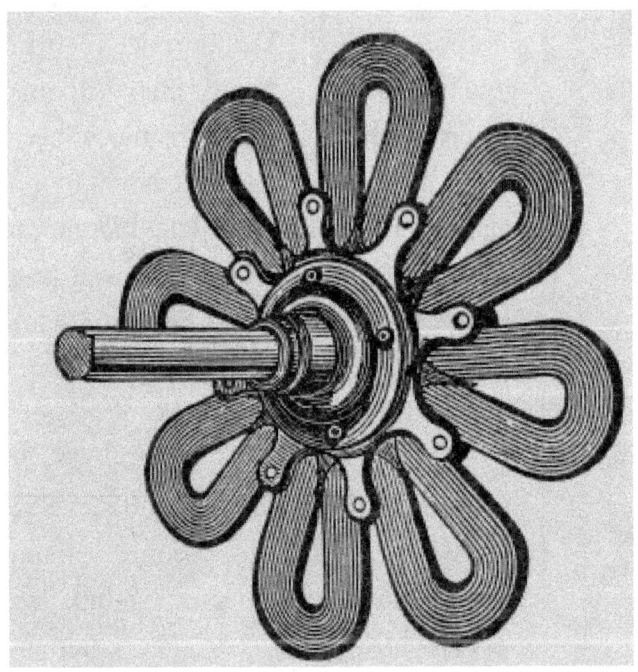

FIG. 9 . FERRANT'S ZIG-ZAG ARMATURE *Source: S.Z. de Ferranti*

The magnetic field was provided by 32 electromagnets whose cores were arranged parallel to the armature spindle and disposed so that the armature was completely surrounded by them:

Ferranti-Thomson Dynamo
(as issued to the press at the time
of its first demonstration in 1882).

Page from S. Z. de Ferranti's
sketch book of 1884
showing early design of meter.

FIG. 10 FERRANTI-THOMPSON DYNAMO. *Source: S.Z. de Ferranti*

The magnetic circuit was completed by two thick iron plates at each end of the electromagnets. The field system was energised separately a small dynamo driven by the same steam engine as the larger dynamo. Two phosphor-bronze bushes collected current from the armature. Early Ferranti-Thompson machines measured only 24" x 20" x

18" and weighted 12 cwt, but supplied over 500 incandescent lamps. Later, larger machines used at the Grosvenor Gallery and Deptford had nominal ratings of 10 000 and 25 000 incandescent lamps respectively.

METERING ELECTRICITY CONSUMPTION

Another of Ferranti's important innovations was his mercury motor meter for measuring consumption of electricity. His early experimental meters were insensitive, but after 1883 he developed a meter which was to receive far wider application. The principal of this meter was the rotation of a pool of current-carrying mercury in a magnetic field. As the mercury rotated it carried a fan with it, which received a torque proportional to the amount of electricity passing through. His early meters were insensitive for three reasons. Firstly, they used an ordinary spiral of copper wire without an iron core to enhance the effect of the field,. Secondly, the air gap between the coil and the metallic ring containing the mercury was too large. Thirdly, the rotation of the mercury pool was retarded by surface tension. Ferranti commented that his early meter required 20 amp to start the mercury

149

rotating. As soon as Ferranti arranged a denser field to thread the mercury pool and isolated the pool from the air, completely immersing the rotor in the mercury, the sensitivity improved to 0.3 amp.

FERRANTI TRANSFORMERS

Ferranti recognized the inefficiency of existing transformers. They used straight magnetic cores and were a source of trouble in distribution systems. So Ferranti decided to produce his own design:

First experimental transformer made by S. Z. de Ferranti, 1885 (*in Science Museum collections*).

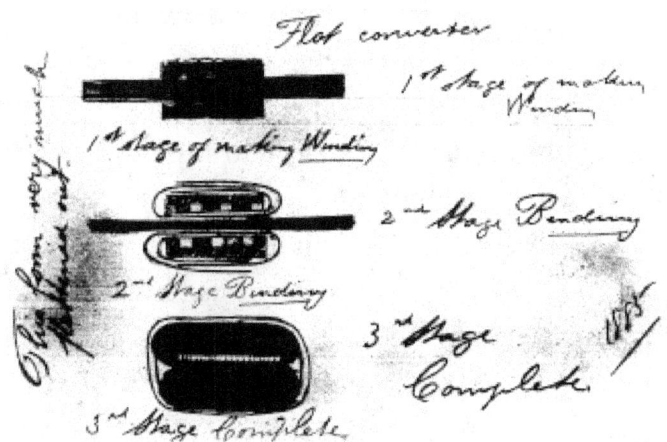

Page from S. Z. de Ferranti's sketch book of 1885 showing original designs for the first Ferranti transformers installed at Grosvenor Gallery and on the Deptford system.

FIG. 11 FERRANTI TRANSFORMER AS USED AT DEPTFORD/GROSVENOR GALLERY

He wound the primary and secondary coil conductors on a closed ring former consisting of many strands of iron wire. Ferranti's transformers were vital components in his high pressure distribution systems which were used later at Deptford generating station.

FERRANTI'S HIGH VOLTAGE SYSTEM OF SUPPLY

Equipped with these developments and his ideas for a high voltage distribution system, Ferranti revolutionised electricity supply. Since late 1883, Ferranti had been manufacturing his alternators, meters and transformers in a small workshop in Hatton garden after the Ferranti-Thomspon and Ince company went into liquidation. His business took him into generating stations and other premises where his equipment was installed, and his assistance was requested by the directors of the Grovensor Gallery Company generating station where the system of distribution was becoming overloaded and inefficient. The Grosvenor Gallery Company station used Siemens 1 000 kW capacity generators at a pressure of 1 200 volt A.C. to light the Governor Street Gallery in Bond Street. The supply was stepped down by straight core transformers connected in series so that the potential drop of each transformer was the same.

Ferranti found the generators were overloaded by the number of consumers connected to it, and the distribution system was inefficient. He temporarily remedied the situation by connecting the Siemens alternators in series to increase the distribution voltage to 2400 volt and used his own transformers, which resulted in a reduced current in the main distribution circuit and a more evenly distributed load. When Ferranti was appointed chief engineer of the Grosvenor Gallery Company in 1866, the existing Siemens alternators were replaced by Ferranti 750 H.P. alternators running at 2 400 volt A.C. , and each with a nominal rating of 10 000 10 c.p. lamps:

Page from Ferranti's sketch book showing design for alternators at the Grosvenor Gallery.

Ferranti 750 h.p. alternators at Grosvenor Gallery, 1887.

FIG. 12 FERRANTI 750 H.P. ALTERNATORS USED AT GROSVENOR GALLERY

DEPTFORD POWER STATION

This system was so successful that a major generating station to ultimately light the whole of London was proposed under a new company, The London Electricity Supply Corporation Limited. The directors looked around for a site for the station where land was cheap and coal and water in good supply. The building was to be erected at Deptford 8 miles from central London on the south bank of the Thames:

153

A contemporary sketch of Deptford from the *Illustrated London News*.

FIG 13 DEPTFORD POWER STATION. Source: S.Z. de Ferranti

Ferranti alternators in the main building were to have an ultimate capacity of 120 000 horse power, but by 1889 when the generating station was opened, generators were in operation of 13 000 horse power capable of supplying over 25 000 lamps. The progress of the venture was beset within many difficulties, not least the legislation of the 1888 amendment to the Electric Lighting Act of 1882 which allocated areas of supply to electric lighting companies. The amendment allocated much of the metropolitan area of London to the Metropolitan Electric Supply Company which was more established in the business with their 6 generating stations. The result was that Deptford

Power Station was allocated a much smaller area than its directors had envisaged.

CONCENTRIC MAINS DISTRIBUTION

Deptford distributed current at 10 000 volt which fed distributing stations where the voltage was stepped down by Ferranti transformers to 2400 volt. The 10 000 volt cables supplying the six distribution stations, known as concentric mains, were laid alongside the south eastern railway line over the Thames into Central London by way of Canon Street station, Ludgate Railway Bridge and Charing Cross bridge. The 2400 volt street distribution mains were terminated in one or more converter areas along the streets, which contained transformers which further stepped down the voltage of the street distribution mains to 50 or 100 volt to supply electric lamps.

This system was the first to operate under such a high maximum voltage, and to employ a generating station some considerable distance outside the load area. Consequently, the general public and many electrical engineers were concerned about the safety and reliability of the system with its array of transformer junction areas. In part, fears were justified, for although the transmission of the high voltage current into the city proved satisfactory and produced negligible loses in power and considerable saving in the cost of copper and other cable materials, the distribution terminal points where voltages were stepped down by transformers were a source of trouble. A serious fire at the Grosvensor Gallery was one result. However, Ferranti's later 10kV transformers worked satisfactorily:

A 10,000 volt transformer from the Trafalgar Square Substation (*in Science Museum collections*).

FIG. 14 FERRANTI 10kV TRANSFORMER.
Source:S.Z. de Ferranti

Ferranti's high voltage A.C. distribution system was soon to be universally adopted and is the basis of today's system of electricity distribution:

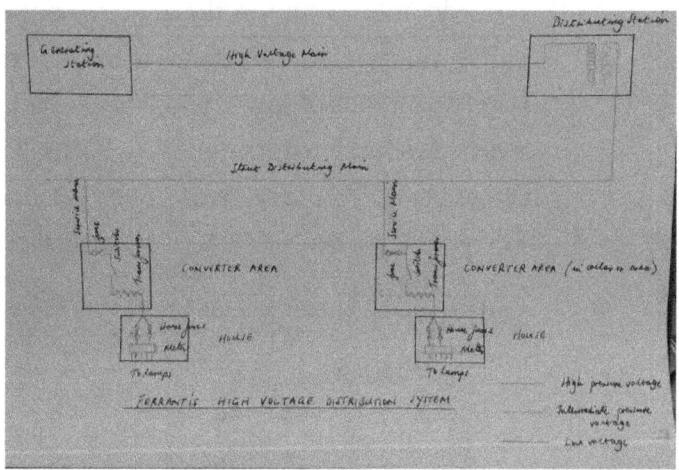

FIG. 15 FERRANTI HIGH VOLTAGE DISTRIBUTION SYSTEM

But the high voltage A.C. system challenged the supremacy of the many small low voltage D.C. installations then in existence. These small interests which were the life's blood of electricity supply all over London were quick to defend their business interests by denouncing the new high voltage system and enumerating the merits of the low tension systems. It was not without a struggle that Ferranti's system won through, a fight which was to become known as 'The Battle of the Systems', which is considered next.

THE BATTLE OF THE SYSTEMS
HIGH VOLTAGE VS LOW VOLTAGE; A.C. VS D.C. SYSTEMS

Reliability Issues

The amendment in 1888 to the Electric Lighting Act of 1882 which increased the period of tenure from 21 to 42 years, and the ease of gaining Provisional Orders, resulted in a firmer foothold for the numerous companies already in existence which were using the low voltage system. The early exploits of the Grosvenor Gallery Company indicated that existing high voltage systems were less reliable and more prone to overloading, mainly because they required the use of transformers which were in their early stages of development. The series arrangement of connection as used in the Grosvenor Gallery Company gave rise to local overloading. Generation of high voltage D.C. using commutators created practical difficulties but until Ferranti's rectifier of 1894 there was no other device for rectifying A.C.:

An early Ferranti rectifier

FIG. 16 FERRANTI RECTIFIER. *Source: S.Z. de Ferranti*

PROVEN LOW VOLTAGE ARC LIGHTING SYSTEMS

As we have seen, arc lamps had been the main source of street lighting for many years before the introduction of the incandescent filament lamps, and direct current was a proven more efficient source for their working. So, it was argued that low voltage systems were desirable for use with street lighting, public buildings, railway stations, etc, where electric lighting was installed.

HIGH VOLTAGE SYSTEMS WITH LOWER LOSSES AND THINNER CABLES

The losses involved in transmitting high voltage current were much lower because the current required is inversely proportional to the voltage for a given transmitted power. Thus, Ferranti's system employed at Deptford's at 10 000 volt required just 1/100th of the current in its power station main to transmit the same amount of power as a low voltage system operating at 100 volt. The thickness of copper used in Ferranti's system could therefore be much reduced, with a considerable saving in the bulk and cost of cables. In considering the diameters of the conductors to use with a given maximum current, the heat generated by the current is the main factor which limits the minimum diameter which can safely be used. The heat generated is given by current^2 x resistance and is therefore proportional to the current squared. In the example given above, Ferranti's system would theoretically require 1/10 000 th of the weight of copper used in low tension systems operating at 100 volt, although this saving was offset by the distance of Deptford Power Station from its load area. Indeed, it was the proximity of low voltage generating stations to their road areas with the associated increase in reliability that formed the main argument of advocates of the low voltage system.

THE CASE FOR SUPERIORITY OF THE HIGH VOLTAGE SYSTEM

To convince advocates of the low voltage system of the superiority of the high voltage system, it was necessary to demonstrate that the reliability of service of the high voltage system was at least equal to that of the low voltage system, and that A.C. could do the work of D.C. equally well. However, before the London Electric Supply Corporation could complete their work on Deptford Power Station, the company was taken to court by Gaulard and Gibbs, who had first used a system of distribution using transformers, for infringement of patent. This case was important in the history of electricity supply because had Gaulard and Gibbs won the case it would have allowed them a monopoly of all forms of distribution using transformers, and progress would have been considerably slowed. Other teething troubles in supply had arisen, among them a fire at the Grosvenor Gallery substation, with the result that the directors were beginning to lose faith in Ferranti's great enterprise. But in his engineers' report of March 1891, Ferranti wrote in defence of his venture:

> *The fact that current of 10 000 volts pressure is transmitted to London every day is the most complete answer to such doubts. The great advantage of the high pressure system is apparent in that the losses involved in the transmission of current from Deptford to the distributing station is inappreciable, while the facilities for procuring coal and water there, sufficiency of room for machinery and appliances, and freedom from the legal and financial consequences attending erection of generating stations in crowded neighbourhoods, cannot fail to tell their own tale in the working expenses of the current year.*

THE BATTLE WON

A turning point in the Battle of the Systems was the decision made by the Electric Lighting Committee to employ the high voltage system to light its streets, public buildings and private dwellings. The lighting in Portsmouth was previously supplied by two separate sources; one source consisted of D.C.

generators running street arc lamps connected in series, the other supplied Swan lamps in private dwellings. The disadvantage of this system was the existence of two distinct lighting systems and the fact that the lighting system feeding factories, shops, etc, was not under load at night. Ferranti produced a rectifier which would enable a single source to supply both public and private lighting by rectifying the A.C. produced by the generators to feed the arc lamps. Four rectifiers, consisting of asynchronous motors with two commutators and a transformer, were incorporated into the Portsmouth electric lighting system. The transformers consisted of a fixed primary coil and a secondary coil which could move away from the primary coil under magnetic repulsion produced by the current load (see Fig.14). The distance between the coils could be regulated by an adjustable weight on the secondary coil, and by this means the current to supply the arc lighting circuits could be kept constant. The great success of Ferranti's high voltage A.C. system at Portsmouth in1894 tipped the balance in favour of the high voltage system which, as Ferranti had prophesied, would become universal:

...in the future our railways will be worked, our lighting will be done, our power will be transmitted to a great distance and this will be done entirely by the aid of High Pressure Electricity; it is high pressure in electricity, like high pressure steam, which is going to carry- which is carrying-everything before it, and this high pressure will be used, and will be doing the work of the world when the low pressure system...has passed away and been forgotten.

THE IMPORTANCE OF THE STEAM TURBINE
Steam Driven vs Hand Driven Generators

As we have seen, the very earliest generators, like those of Hippolyte Pixii, were driven by a handle, but as machines became larger to produce more useful current, it became obvious that an engine capable of supplying more driving force for longer than was possible with human muscle was necessary.

THE RECIPROCATING STEAM ENGINE

The English engineer Frederick Holmes was the first to apply the steam engine to the electricity generators to energise arc lamps. This generator, with an output of $1^1/_2$ kW, was driven by a reciprocating steam engine demonstrated in London on 1857. Based mainly on the pioneering work of James Watt and Trevithick, steam engines of this period after 200 years of innovations and improvements were still basically the same as the earliest piston and cylinder designs. But as machines, materials and manufacturing techniques improved, so did the reliability and efficiency of stream engines. Generators were almost exclusively rotated by reciprocating steam engines.

Crompton used a steam engine in the 1870s very similar in design to the Trevithick engine to drive Gramme generators which were used in his generating sets. The steam engine and generator were connected via a pulley belt (Fig. 7). The same method of connection was used even as late as 1889 to couple his Hargreaves 1 500 hp steam engines to Ferranti 1 000 kW alternators at Deptford.

POWER ABSORPTION OF BELTING AND SHAFTING

But a considerable amount of energy was absorbed by inefficient belting and shafting that transferred power from the steam engine to the generator armature.

Direct coupling and Vibration Problems

It was Edison who proposed to couple a steam engine directly to the armature of his bipolar dynamo to be used ultimately at his Pearl Street Station. The armature and steam engine previously mentioned designed by Sims were mounted on the same cast iron bed plate, but Edison was to suffer many problems in the running of this arrangement due to vibration produced by the reciprocating motion of the steam engine, which could not be perfectly dynamically balanced as the component parts were not precisely engineered.

FLYWHEEL DAMPING OF VIBRATIONS

The Hick and Hargreves engines used at Deptford contained an improved valve gear invented by G.H. Corliss in America, high and low pressure cylinders were coupled in tandem and a 390 ft flywheel was used to dampen torque fluctuations produced by the engine running at 300 r.p.m.

DOUBLE AND TRIPLE EXPANSION ENGINES

A most important advance of this period was made by P.W. Williams who took out patents in 1884 and 1885. He developed high pressure engine with multiple cyclinders supplied in a system of vertical piston valves working up and down inside the piston rod which had ports cut in it. More power was extracted from the steam in two and later three stages. All parts were machined to a tolerance of 1 thou. Many of these engines were in operation between 1890 and 1900. Six Williams 150 HP steam engines directly coupled to Crompton generators were used for lighting the Opera House, theatre and public buildings in Vienna in 1889. The generators, supplying over 700 kW at a pressure of 100 volt between legs of his five wire system were connected across lead acid batteries . When started in 1885 this was the most ambitious central generating station project then attempted. The great success of Crompton's system led him to form the Kensington Court Company which installed 7 Williams steam engines driving Crompton generators at Kensington Court to supply over 100 large houses on a new estate.

THE CONTINUING INEFFICIENCY OF STEAM ENGINES

Even after the innovations of Williams and Corliss, the reciprocating steam engine was measured to convert less than 12% of the available energy of coal into mechanical power. In the 1890's when electricity supply was responsible for transmitting vast quantities of energy, a prime mover with such a low efficiency involved a great wastage of power. Furthermore, with the universal adoption of high voltage systems of distribution, alternators had to be rotated at high speed to produce a high e.m.f. with high output. The solutions these problems lay in the development of the steam turbine.

THE STEAM TURBINE: THE WORK OF CHARLES PARSONS

When Charles Parsons took up the idea of developing the steam turbine in the 1880's, the principle involved was an old one. Hero's 'Sphere of Aeolus' of the first centre A.D. used a jet of stream to turn a vaned wheel, and was the first reaction turbine. Trevithick's 'whirling engine' consisted of a pair of hollow arms mounted on shaft. At the end of each arm was a small hole from which steam at a pressure of 100 p.s.i. emerged causing them to be driven around. This machine could only rotate at 250 r.p.m. and therefore could harness only a fraction of the energy of the steam. At the age of 22, Parsons took an apprenticeship at an engineering works in Newcastle upon Tyne. He envisaged a wide range of applications for the steam turbine, the most important being as a prime mover for generators of the rapidly rising electricity supply industry. Parsons was convinced that larger, central generating stations would come into existence supplying much larger load areas and requiring larger engines to drive the generators. Parsons set to work in developing a steam turbine which would run reliably at much higher speed and with high efficiency than piston engines.

THE AXIAL FLOW TURBINE

Parsons was aware that high pressure steam travelled at thousands of f.p.s., so that to harness the energy of the steam the turbine would have to run at a comparable speed. At this speed, centrifugal forces would be tremendous and would make the machine almost unworkable with the technology off the day. He had to find a way of making the steam deliver nearly all of its energy to the turbine at a much lower speed. His idea was to make the steam expand in stages by passing it through a series of elemental turbines. His axial flow turbine consisted of a fixed circular casing, the stator, with rows of stationary blades on the inner surface. A shaft containing a ring of similar blades, the rotor, rotated within the stator. Steam flowing axially through the rotor had to pass between the rings of the blades, each of which acted as a nozzle through which the expansion of the steam could take place in stages, driving the rotor round.

Parson's axial steam turbine was a most efficient engine, and after taking out patents in 1884, he coupled his turbine to a small generator running at 18 000 r.p.m. which produced 75 amps at 100 volts.

After a successful demonstration of a portable 4 kW turbogenerator used for lighting a public event at Gateshead, the demand for Parson's 'turbogenerators' grew and the Newcastle and District Lighting Company was founded by him in 1887 to produce bigger turbogenerators. In the December of 1883, after the properties of low steam consumption, space saving, increased reliability and freedom from vibration of the turbo generator had been proven, the City Corporation of Elberfield in Industrial Rhineland ordered two 1 000 kW turbogenerator sets.

Interest in Parsons turbogenerators in England was also increasing. At Portsmouth in 1894, a Parsons 150 kW turbogenerator running at 3 000 r.p.m. was installed alongside conventionally driven Ferranti alternators running at 96 r.p.m. as part of a new high voltage system installed by Ferranti. Ferranti had been interested in the steam turbine for some years, but the efficiency and high speed smooth running of a Parson's steam turbo alternator running alongside his own slow speed alternators brought his interest to fruition.

Thereafter, Ferranti began many years of experimenting with Vickers River Dan Works at Sheffield. He produced his first turbo alternators in 1895, an early application of which was at the Hammersmith Generation Station where 125 kW alternators were employed in 1896.

Variations in temperature, differential rates of expansion, leakage and other mechanical problems tended to reduce the efficiency, reliability and power output of the early steam engine. Ferranti, after many experiments, found techniques of manufacture and materials which overcame these problems. His later turbines contained rotors whose blades consisted of accurately finished mild steel with a thin coating of pure sheet nickel electrically welded on to the surface. To prevent loosening of the rotor blades at high temperature, caulking between the blades was used which expanded during operation. The blading was also electrically welded to the rotor shaft.

STEAM REHEATING

Another innovation of Ferranti's was his steam reheating and bleeder heating. Before expansion the steam was superheated and after the first expansion, the steam was re-superheated and passed on to another stage of the turbine. After this the steam was exhausted into a regenerator and then a condenser. Parson's steam engine, with the addition of many innovations invented by men such as Ferranti, is still the principle prime mover used to rotate the great generators of today. From the first $1^1/_2$ kW generator driven by a crude reciprocating engine in 1857 have evolved the massive 500 000 kW turbo-alternators of modern times.

FIG. 17 FERRANTI STEAM TURBO-GENERATORS. *Source: S.Z. de Ferranti*

But cables to carry the increasing output of such generators required suitable conductors and insulation. Power cables and the distribution of electricity are considered in the next chapter.

8. THE TRANSMISSION OF ELECTRICITY
EARLY CABLES

The electric power cable owes much of its early development to the search in the late 19th century and early 20th century scientists to find a satisfactory means of transmitting telegraph signals. The earliest cable for this purpose merely consisted of bar copper conductors separated by air installed as overhead lines.

SAVING SPACE; SOLID INSULATION

This arrangement occupied a great deal of space and it was suggested that, as only a small gap between conductors was necessary, a large number of conductors could be arranged together in the same cable, separated by a solid insulating material.

INSULATING BURIED CABLE; PAPER AND COTTON

An early material used for this purpose was paper, applied over each conductor and coated with pitch, the whole then being covered with more paper.

Telegraph engineers were quick to see the advantages of better protection of the underground system and a new form of insulation, cotton, was tried by Wheatstone in 1837 as a covering for conductors laid in wooden baulks covered with wooden strips and painted with preservative. This arrangement did not preserve the insulation, but when iron pipes were used instead of wooden baulks a more satisfactory system resulted. Additional protection was afforded by the use of lead sheathing in 1846 and later in 1854 steel tape armouring was introduced to protect cables against mechanical damage.

The use of cotton for covering electric cables had the disadvantage that, in an unimpregnated state, its hollow fibres acted as capilliary tubes which drew in moisture. This, in addition to the electrolytes dissolved in the cotton remaining from the manufacturing process, gave rise to an undesirably small insulation resistance between adjacent conductors. The textile material also degraded quickly. The use of cotton as an insulator was important, however, because the search for an efficient means of manufacturing cotton covered conductors led to collaboration between textile engineers and telegraph engineers in adapting machines previously used for manufacturing cotton braid and cords to apply cotton braid over conductors. The adaptation of such machines around 1875 led to considerable output of cotton covered cables supplying the telegraph industry mainly by Salford Electrical Wire Works formed in 1868, and W.T. Healy manufacturers of silk and cotton covered wires.

GUTTA-PERCHA INSULATED CABLES

The first use of electric power cables was in the lighting of lighthouses and then, more significantly, the lighting of the West India Dock and later Billingsgate Market, Holborn Viaduct and part of the Thames Embankment with Jablockhoff lamps in 1877 and 1888. The cables used consisted of seven strands of 16 gauge (0.064") conductors covered with gutta-percha insulation and tape and run in troughs underground, which were still functioning 5 years later when the West India Dock installation was decommissioned.

Gutta percha was introduced into Europe in 1843 but its desirability as an electrical insulator had been discovered by Faraday. Companies first came into existence to produce gutta percha cables for telegraph lines and subsequently for power cables. It is a natural thermoplastic material which is easily shaped whilst hot, and when cold it has a hard, leathery consistency. It is a gum occurring in trees of the family *Sapoptacae* found in Malaya and the East Indies. Gutta percha has a similar structure to natural rubber, but it has a harder consistency when cold. It has a high insulation resistance (about 10^{15} Ωcm^{-1}) and A.C. losses due to heating are also small. Unlike many other insulants, it is capable of retaining a considerable proportion of moisture without a serious effect on insulation. However, it could not be used on a large scale because of its high cost and instability at high temperatures.

OVERHEAD SYSTEMS

An alternative to insulating cables with some form of dielectric covering was, of course, to employ uncovered conductors in overhead systems. Kings Cross station, with its high roof, was well suited to overhead installation which was adopted for its lighting system in 1882. An engine room containing 5 generators was erected at the station, each supplying 3 arc lamps fed by copper strips mounted on porcelain insulators. Crompton's copper strip distribution system was employed later at a much larger public supply generating station to supply subscribers at Kensington Court in 1886. The main conductors were 1" thick operating at 200 volt. This system had the advantage that additional strips could be laid on top of existing strips to increase load capacity. However, Crompton's strip distribution system was naturally only usable at subscribers' premises suited to the installation of overhead wires. The majority of cables distributing electric power had to be laid in the ground directly, and as such were liable to more rapid degradation, or in troughs or pipes which was much more expensive. Gutta percha insulation, applied by a machine invented by Hancock in 1848 in a continuous coating over conductors, was not very satisfactory as it was liable to perish. Vulcanised Gutta percha offered better resistance to corrosion, but its sulphur content chemically attacked the copper conductors it was insulating. Edison's distribution systems in America, for example in Pearl Street and Holborn Viaduct in London, were early examples of the use of bitumen for insulating power cables, considered next.

BITUMEN INSULATION

Bitumen both as a form of insulation and a barrier to moisture was more successful than gutta percha owing to its resistance to chemical attack. A compound of bitumen, known as vulcanised bitumen, was introduced by William Callender who imported Trinidadian bitumen for use in road surfacing. It consisted of a mixture of 20 % bitumen, mixed with sulphur and Elastikon, a pitch residue from the manufacture of cotton seed oil. Callender felt assured of the potential of the compound as an insulator and formed the Callender Bitumen Telegraph and Waterproof Company in 1882. The company covered their conductors with jute serving and laid them in iron or wooden troughs fitted with a cover. The trough was then filled with vulcanised bitumen. This, known as the Collender Solid System, had the advantage that sections could easily be jointed by slipping ferrules made of copper strip over each end of the conductors to be joined and then soldered in position. The vulcanised bitumen compound was then introduced to insulate the joint. By the end of the century, the use of bituminous compounds as insulants was widespread.

INDIA RUBBER INSULATION

Vulcanised India rubber, introduced in this country in 1859, gained a reputation as the best power cable insulant until the turn of them century when the superiority of paper insulation for use in high voltage cables had been proven. Natural rubber consists of latex- a milky liquid of certain species of tree- which has been coagulated by acids such as formic and acetic. Purified rubber has a high insulation resistance but absorbs moisture and deteriorates by oxidation. Pure rubber heated with sulphur to produce vulcanised rubber, is much more stable. Vulcanised rubber has a high tensile strength coupled with high elasticity and therefore made an excellent insulant for use at junctions in service connections where flexibility was required. As with vulcanised gutta percha insulation, the sulphur in vulcanised rubber tended to corrode copper conductors, but this problem was overcome by tinning the conductors. An early application of vulcanized rubber insulation in power cables was in the Grosvenor Gallery Distribution system in 1884. The cables were supplied by the India Rubber, Gutta Percha and Telegraph Works Company in Essex, and consisted of 19/15 gauge tinned copper wires insulated with 0.21" vulcanised rubber applied in tube form. The system operated at 2 400 volt; the cables were distributed from a lattice tower on the roof of the gallery, and supported from poles 3 3/4" diameter in cast iron sockets on the house tops. The cables were suspended by leather thongs 7" long and 3/4" wide from steel cables, which were shackled off at each pole. As electricity supply became more ambitious, the lengths and sizes of cables used to carry electricity from generating station to consumer increased greatly. However, if vulcanised rubber insulation was to be used in high voltage central generating station installations the amount of rubber required would be great and the cost very high.

IMPROVEMENTS TO PAPER INSULATION

Paper insulation was cheap and its mechanical and electrical properties could be controlled more easily than rubber, but paper tended to absorb a considerable amount of moisture. The solution to this problem began the development of an insulating material in cables successfully operating at 100's of kV in modern day systems.

EDISON'S DISTRIBUTION SYSTEMS

In his Menlo Park laboratory in 1880, Edison experimented with new schemes of laying wires to produce an arrangement which would not produce local fall in voltage but which would save copper. We have seen that Edison designed a new system in which wires were to be laid along each street and connected at each end in junction boxes, which he called mains. Feeder wires would be connected to the mains at one end and the source at the other. This was the distribution system Edison installed at Christie Street as a demonstration to financiers of the future Pearl Street central generating station.

Trenches were dug along Christie Street, and the mains consisting of copper conductors run along grooved planks. Not surprisingly, current leaked badly from the conductors. His proposed system still had to be proven and Edison's workers dug trenches along Christie Street ready to accept the distribution conductors. Edison was venturing into unknown territory, and he was faced with the problem of deciding how to lay his cables unaided. Edison soon discovered the need for good insulation. His conductors and planks had to be dug up because of leakage of current through electrolytic action in the damp earth. It was clear that conductors would have to be insulated with a substance which would give a high resistance to earth and prevent the ingress of moisture.

EXPERIMENTS WITH DIFFERENT INSULATION

There followed many experiments with conductors in situ using different insulating materials. An early attempt to improve leakage was made by pouring coal tar over the wires, but this too failed. A study of the limited literature on the subject of insulators then available led to the adoption of a mixture of Trinidad bitumen, paraffin and beeswax. Barrels of linseed oil, bales of cheap muslin, tons of bitumen and two large iron kettles were purchased and the muslin strips passed through the hot mixtures and wound into balls. The conductors were supported on wooden saw horses and three boys each equipped with a ball of impregnated muslin tape ten feet apart straddled the trench and wound the tape spirally around the conductors. The mains thus used consisted of a ten gauge copper conductor insulated with spirally-applied bitumen-impregnated muslin tapes. When laid in the ground the mains insulated in this manner did not leak current. The 'feeders' and 'mains' laid as part of the pearl Street supply system were to be free from leakage for some years without deterioration, and cables were thereby permanently insulated and protected form their environment.

CORROSION PROTECTION; EDISONS TUBES

Edison conductors laid in iron pipes became known as 'Edison tubes'. The mains consisted two large symmetrical shaped copper conductors separated by cardboard and suspended on jute fibres in 20 ft long Edison tubes; the maximum length that could be transported to its destination along the narrow New York City streets. The pipe was permanently insulated with bitumen using a suction pump to draw in the bitumen. By the late autumn of 1881, some 5 miles of mains had been laid in Manhattan to carry low voltage D.C. of 100 volt. Trinidad bitumen, or Trinidad Lake Asphalt is found in its natural state as a mixture of hydrocarbons and about 40% by weight of mineral matter. The use of Trinidad bitumen for Edison's distribution cables was very satisfactory in isolating the copper conductors from their local environment, bitumen having impressive impermeability to moisture and insusceptibility to dilute acids found in soils. Its electrical properties, however, such as specific insulation resistance are poor, but the thickness used and the low voltage which the conductors carried allowed their use to give a satisfactorily insulated and reliable cable. One could well believe, however, that the Edison New York City Lighting company faced subsequent leakage problems due to the formation of voids in the insulating material because of the poor flow properties of bitumen. Ensuring uniform adhesion to the conductors would also have been a problem. The most that the company could do under the circumstances was to have an engineer check each section of the cable for insulation resistance as it was completed, which was tested using a galvanometer.

EDISON'S THREE WIRE SYSTEM

Shortly after the opening of the Peal Street station, Edison tried out a new idea that would, if it proved satisfactory, enable a more economical use of copper in distributing electricity. To test his idea, Edison stretched three parallel wires on a board, with two rows of lamps connected between two outer wires and a middle one:

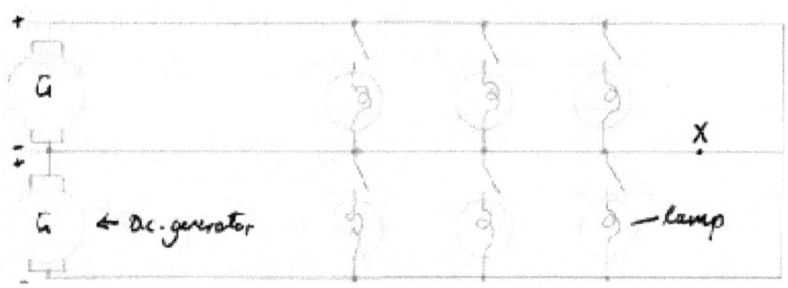

Edison's Three Wire System

FIG 18 . EDISON'S THREE WIRE SYSTEM

Two of his D.C.dynamos were connected in series been the outer wires, the middle wire being connected to the junction between the dynamos. Edison was especially interested in the effects of loading between the outside wires and the middle wire on the current carried in the inner wire. To test this, Edison installed a German silver with known resistance at X and connected a galvanometer in the circuit to obtain an indication of the current in the middle wire while he switched the lamps on and off. He found that the middle wire carried the out of balance current of the two sides of the system, so that when the number of lamps burning in each row were equal, the middle wire carried no current. Therefore the middle wire could be made much thinner than the outer wires by balancing the loads in the two circuits. This system, also invented independently in 1882 in England by Hopkinson, is called the three wire D.C. system for low tension distribution. Edison had correctly predicted that only about 1/3 of the weight of copper used in the two wire system would be required in an equivalent three wire system.

Edison intended to prove the worth of this three wire system by installing it at Brockton, Massachusetts and at Sunbury, Pennsylvania, both locations strategically chosen because of cheap coal but expensive gas. However, the residents of Sunbury objected to the scheme, some saying that the current would 'leap from the wires and burn their house down'. But the 3 wire system of distribution worked exactly according to Edison's plans:

> *I have come to the conclusion that my system of lighting, after being perfected, should be promoted..*

The Edison Electric Lighting Company was also active in London. Shortly after the opening of Pearl

Street station, the Holborn Viaduct installation was put into operation in January 1882. The central generating station was erected at No 57 Holborn Viaduct and the Edison tube rigid main system was run in subways supplying street lights and public buildings from Newgate Street to Holborn Circus. In 1883, the mains were modified to work on the three wire system, the Edison tubes now containing three copper conductors arranged in trefoil formation, each being individually lapped with jute cord. The conductors were insulated with bitumen as before. However, as voltages increased with increasing demand, new problems with materials and in the design of cables came to the fore. The problems encountered in early high voltage cables are considered in the next section.

EARLY HIGH VOLTAGE CABLES

MOISTURE PROBLEMS IN PAPER INSULATED CABLES

Early papers for insulation of power cables were made from abaca trees found in the Philippines. The cellulose fibres of the leaves , manila fibres, were deposited on a grid to make manila paper. The finished product contained non-cellulose cell substances associated with the fibres, mineral matter absorbed from the soil and natural gum and silex residues. The presence of these impurities gave rise to electrical properties far below the potential of pure paper, but until the advent of high voltage distribution, this merited less consideration than the mechanical properties required for use in machines applying paper lapping.

OIL IMPREGNATED PAPER INSULATION

The most immediate problem to be overcome was how to dry out the moisture already present in the paper insulant, and how to prevent the latter from reabsorbing moisture in use. Early impregnated paper insulated cables consisted of conductors insulated by paper tape wound spirally around them. Impregnation was carried out by immersing the insulation in the hot oil or other impregnating compound, hoping that all the moisture would be driven out by the heat and that the interstercies in the fibrous material of the paper would be completely filled with the impregnating compound. A more effective process was invented by Jaques in America, described in his patent of 1885:

The process of insulating an electrical conductor consists of covering the conductor with fibrous insulating material, subjecting it to the action of heat and an air pump operated to produce a vacuum until the combined action thereof bursts the cellulose structure of the fibrous insulating material and dispels the moisture, and then applying the insulating material under pressure

TECHNIQUES FOR APPLYING PAPER INSULATION

The techniques for applying gaper insulation were derived from the rope making and textile industries, and as early as 1850 paper lapping machinery had been invented in which the cable was rotated past stationary spools of paper or the spools rotated about the cable as it moved past them, so that a helical lapping of insulation with the desired number of layers was obtained. The latter type of insulation was developed by American McCracken in 1886, and its success in applying paper lapping to conductors led to the founding of the Norwich Insulated Wire Company. The cables produced consisted of conductors insulated with 3/4" wide manila tape lapped upon itself by half the width of the strip, and these were spirally wound together in cores. A lead sheath was extruded onto the cable and the insulation impregnated with hot oil. Although the Norwich Insulated Wire Company specialised in producing telephone cables, the process used in paper lapping, assembling the cores or 'laying up' and impregnating could be applied directly to the manufacture of power cables. After Ferranti had shown the superiority of impregnated paper insulation in high voltage cables, an alliance between the directors of the Norwich Insulated Wire Company and Ferranti resulted in the manufacture of impregnated paper insulated power cables.

FERRANTI'S DEPTFORD POWER CABLES

When the Deptford power station opened in 1889, the use of paper insulated mains to transmit a current at 10 kV form Deptford to London everyday showed the practicability of paper insulation:

Ferraanti High Voltage Paper-Insulated Cable
Top: 10 kV rigid main 1889
Left: Two experimental lenghts 1893
Bottom Right: New Deptford Main

PLATE 14: FERRANTI HIGH VOLTAGE CABLES. *Source: S.Z. de Ferranti*

In deciding which type of cable to use for this purpose, Ferranti could not draw on the experience of others working with high voltage distributions-there wasn't any. The gutta percha, rubber and jute insulated cables then manufactured had not been operated at pressures above a few hundred volts. Ferranti used the results of the voltage required to produce sparks of certain lengths between spheres in air, produced by Kelvin, to estimate the thicknesses of insulation required to provide adequate insulation between conductors. He found that in the case of gutta percha and rubber, the thickness of insulation required would make cables prohibitively expensive, whereas with jute, there was a fire risk due to the temperature of operation of the cable.

If Ferranti wanted to use a voltage 50x greater than any one else had done to distribute electric current, he would have to design his own special cables to carry it, just as he had had to design his own generators, transformers and switch gear for use at such a high voltage. The cables designed for Deptford by Ferranti used copper tubes as conductors. The inner conductor of $1^3/_{16}$" diameter was drawn into an outer conductor of the same cross-sectional area 31/16" in diameter, the space between being filled with dielectric material. For the latter, Ferranti used wax-impregnated paper. The decision to use this dielectric came through its use in capacitors, being capable of withstanding high electric stresses, and because of its availability and the cheapness of commercial paper. Also of great importance was the existence of machinery and the techniques for applying paper insulation. It is interesting to note that the choice of 10 kV as a voltage for Ferranti's distribution system appears to have had little scientific basis other than some application of Kelvin's table of sparking potentials.

FERRANTI'S EXPERIMENTS ON INSULATING BREAKDOWN VOLTAGES

Ferranti did no experiments to ascertain maximum electric stress that various samples of commercial paper or impregnate would be capable of withstanding without breaking down, or on the ageing characteristic of insulants, or the effects of cyclic loads and expansion and contraction of cable materials which gave rise to voids and impaired insulation resistance. In fact, had Ferranti chosen a much higher voltage, such as 33 kV, as was adopted around 1920 due to vastly increased power loads, he would have come up against problems such as disruptive failure, dielectric deterioration due to gaseous ionization, and thermal breakdown, which have only been solved relatively recently in the light of scientific knowledge and techniques which were not available to Ferrant at the time. In the Ferranti cables, wax impregnated paper 36" wide from sheets of 20 ft in length was rolled spirally by machine onto the inner conductor until an insulation of 1/2" thickness was obtained. The outer conductor was then extruded onto the insulated inner conductor.

INTERFERENCE BETWEEN POWER AND TELEPHONE CABLES

The cable in this condition was still, however, unsatisfactory. Engineers of the Post Office would not approve Ferranti's concentric mains because it would produce inductive effects in telephone cables in the vicinity. After a series of experiments on the cable, a solution to the problem was found. The outer conductor was covered with a conducting lead sheath connected to earth. This meant that the cavity between the outer conductor and the sheath contained no electric currents and therefore no field, so that no interference with close-by telephone circuits was noticeable. The fully developed Ferranti main consisted of two concentric tube conductors with 1/2" insulation, a further 3/32" of wax impregnated insulation over the outer conductor and a thin iron tube of 2 3/8" diameter enclosing the whole. The cable was manufactured in 20 ft lengths

CABLE JOINTING

To make connections between lengths, Ferranti had to design a simple joint since the four seven mile mains to the substations in London required over 7 000 of them. The jointing was done by turning down one end to be joined to a male cone, and the other end to a similarly shaped socket. The inner conductor of each end was reamed out to accept a solid copper rod pushed in at both ends. A copper sleeve was pressed tightly onto the outer conductor at each end by means of a special tool after drawing the two ends together with a jack. An iron sleeve of the same inside diameter as the outer iron pipe was drawn down over the joint and insulated with a paper sleeve. The joint was then filled with molten wax under pressure.

FAULT FINDING

Test boxes were employed at intervals along the mains, in order that faults in sections could be located .These consisted of metal boxes into which the two ends of the cable were drawn and the conductors joined with clamps and bus bars which could easily be disconnected. Faults were located using a Wheatstone bridge circuit employing a galvanometer.

After the Deptford main was completed in 1891, Ferranti joined forces with the Norwich Insulated Wire Company who went into production of 11 kV paper insulated concentric cables. The outer conductors comprised flat segmental strips and a lead sheath applied with serving and, where required, steel armouring tape. The 11 kV cable had still to establish itself, however, and the first 11 kV cables produced were little more than prototypes. The real business for the company was in the supply of low tension cables of 2 000 volt or less in systems using low voltages, though, which were still very common. A typical specification for a 11 kV cable produced by the company was 7/0.064" diameter inner cores giving as cross-sectional area of 0.0225 in^2. Such a cable was used as an experimental replacement for a Deptford main and worked very well. In 1898 a 42 mile length of 10 kV flexible concentric cable with 0.54

in² conductors capable of carrying 7.5 MVA per phase was produced for the Metropolitan Electric Supply Company. By 1900, the high voltage impregnated paper insulated cable had firmly established itself as the cheapest and most efficient means of transmitting large amounts of electric power. But further improvements in economy and efficiency were to come with the advent of *3 -phase* distribution, which is considered next.

THREE PHASE ALTERNATING CURRENT SYSTEMS: WOOD LANE 1900

An event of considerable importance to the future economical use of cable materials and reliability of cables in service, was the opening of a power station at Wood Lane in Hammersmith in 1900. The coils of the generators at the station were connected together so that *three* wires carried the current to the mains. Each wire carried a sinusoidal voltage which differed by 120° or 240° in phase from the other two. The Norwich Company supplied seven miles of 3-core 5 000 volt cable for distributing the current. Each core carried one phase which varied from the voltages in the other two cores according to their vectorial sum, as explained in Appendix I.

FEATURES OF THE 3-PHASE SYSTEM

The important feature of this system was that the summation of the voltages in the three cores is zero, so that three wires only were necessary to carry the current instead of the usual six, provided the load was symmetrical. A very considerable saving is therefore achieved. The arrangement of the coils connected in the generator is shown below:

FIG. 19 ARRANGEMENT OF COILS IN A THREE-PHASE GENERATOR

BALANCED LOADS

In order to demonstrate the principle only three wires are required to carry the current in three phase systems with symmetrical or balanced loads, the diagram below has been drawn:

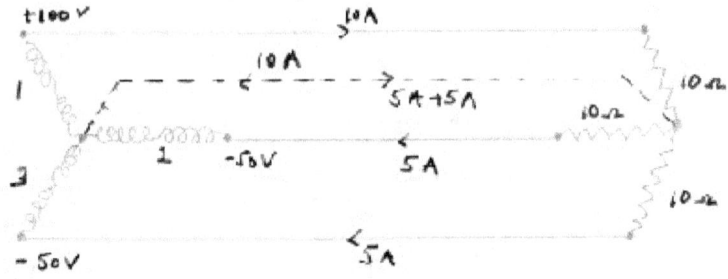

FIG. 20 A 3-PHASE SYSTEM WITH BALANCED OR SYMMETRICAL LOADS

The diagram shows a three phase generator with balanced resistive loads of 10 Ω. From the diagram, taking the common point of connection of all three coil circuits as 0 volt and the potential of the top wire as 100 volt, say, the potentials of the two lower wires is -50 volt. If a fourth wire is connected, a current of 10 A passes from coil circuit 1 clockwise, but this would be balanced by 5 A from coil circuit 3 anticlockwise and 5 A from coil circuit 2 anticlockwise. As a result, the fourth wire carries no current and so is not required. Although a fourth wire is not required with balanced loads, the cores need to be fully insulated from earth, because if a fault develops in one core due to low earth resistance, the voltage of the other two lines w.r.t. earth increases to the same value as the voltage between lines, which in the event of inadequate insulation between cores and earth breaks down the cable.

THE SPREAD OF 3 PHASES A.C. SYSTEMS

Since 1900, the use of three phase A.C. has become universal, but to cope with the increased demand for electricity from industry and domestic consumers, the high voltage used for distribution has been greatly increased, bringing with it new problems of design. These problems are considered in the next section.

SOME PROBLEMS IN E.H.V. CABLE DESIGN INCREASING SUPPLY VOLTAGES

Since the early historic cable installation by Ferranti at Deptford in 1891, engineers have been continually challenged by the rapid growth of power load and the increasing necessity for a cable capable of operating satisfactorily at higher and higher voltages. The introduction of 3 phase distribution in 1900 represented a great economy materials and simplified cable manufacture, but the rapid increase in demand from the electrochemical industry and domestic lighting, heating and electrical appliances necessitated the introduction in 1920 of 33 kV cables. At this voltage, problems arose which could not be solved simply increasing the thickness of insulation alone.

PROBLEMS OF 'OHMIC' HEATING

There were other problems too. For the above reasons, it was necessary to excavate land on an increasingly large scale to lay cables underground or drawn into pipes or ducts. Besides this, the current carrying capacity of underground cables was limited by the heat evolved during transmission of power. The heating effect is produced by the ohmic resistance of the conductors, and the energy supplied to the capacitance of the cable to charge it. The fact that underground cables had insulation around the conductors, lead sheathing, armour and often bedding and serving, meant that the rate at which heat could be conducted away from the cable into its surroundings was restricted. In addition, the environment consisted typically of around 4 ft of soil above the cable, and soil is a poor conductor of heat. Hence the current carrying capacity of the cable had to be kept below some critical limit at which the heat generated exceeded the rate at which it could be conducted away. Above this limit, the temperature of the cable would continue to rise until the cable materials broke down thermally.

E.H.T. DISTRIBUTION IN OPEN COUNTRY

This problem was circumvented in open country by supporting bare conductors on steel towers ('pylons'). The cables were insulated from the towers by glass or porcelain insulators and large air spaces. The current carrying capacity of a conductor of a given cross-section is considerably increased, since the heat generated can be directly communicated to the surrounding air. Installation costs per mile of cable operating at a voltage of 132 kV in 1962 were 7x more for underground cables than overhead cables, and greater still today. Obviously, though, in cities overhead distribution cannot be used on a large scale and therefore underground cables have to be used.

PROBLEMS WITH EARLY 33 kV CABLES AN THEIR SOLUTION

Early 33 kV cables gave unsatisfactory service for several reasons. The first was that the high voltage created a high density electric field in the dielectric material of their insulation which forced the impregnating material to flow under stress, creating spaces in the dielectric known as 'voids'. Gaseous ionization of the air in the voids then took place in which the impregnating compounds were decomposed into carbon, while deposits tended to spread between the layers of dielectric, leading to ultimate breakdown. It was clear that this problem would become more acute in the future as cable voltage increased, and therefore further advances in the technology of power transmission rested on an effective solution to this problem. This was found for 33 kV cables by surrounding the cable cores with metal foil to ensure a uniform radial field in which there were no concentrations of electrical stress. But this innovation was found unsatisfactory when higher voltages were used.

PRESSURIZATION; OIL IMPREGNATED CABLES

A better solution was to apply a mechanical force to counteract any stress within the cables. This was done by applying pressure to the cable insulation and sheath by pressurised gas or insulating oil. The oil filled cable uses a low viscosity mineral oil for impregnating the dielectric in channels within or adjacent to the insulation, so that when pressure is continuously applied to the oil , no void formation occurs. Pressures involved are around 100 p.s.i..

GAS FILLED CABLES

The gas-filled cable uses a pressure of 200 p.s.i. nitrogen which compresses the dielectric via a thin lead diaphragm. Expansion and contraction of the dielectric impregnating compound is controlled by the high gas pressure.

Another reason for unsatisfactory early operation of E.H.V.cables was thermal instability. As previously noted, the maximum loading of cables had to be below some critical value, and so increases in power capacity had to come via the use of higher voltages.

MODERN E.H.V. DISTRIBUTION

Today, generating stations are linked by overhead lines operating at 132, 275 or 400 kV to an electrical control network which feeds substations within supply areas. Heavy feeders supply feeding centres which in turn supply adjacent streets via distribution pillars. The stages of distributing electric power from generating station to consumer bear a striking resemblance to Ferranti's method of distribution from Deptford. Perhaps even Ferranti did not envisage a national network operating in some sections at nearly half a *million* volt and transmitting vast quantities of current, but he strongly believed that the future of electricity lay in the utilization of higher and higher voltages. The present trend in increasing voltages cannot go on indefinitely however with the same materials used in the same configurations. There is a limit to the electric stress which can be tolerated by impregnated paper insulation, but no doubt synthetic dielectrics will come to the rescue in enabling still higher voltages to be used.

9. CONCLUSION

The study of electricity and magnetism arose from the need of the ancients to find explanations for the forces of nature. The properties of electrostatically charged objects and magnets were spectacular, and in the absence of scientific knowledge it is not surprising that magical interpretations of their behaviour and interpretations based on mythology were given. But the efforts of the ancient Greeks were important because in order to explain natural forces, they had to find out more about their behaviour and thence came attempts to experiment with charged bodies and magnets in the hope that investigation of their properties would reveal explanations. The observations of the ancient Greeks and Chinese inspired the interest of later research by Peregrine and Gilbert who studied electrostatic and magnetic phenomena more systematically, carefully noting their observations and findings, using experimentally determined facts to back up their theories.

The treatises of Peregrine and Gilbert, were, then, early works which used scientific method and as such were unbiased attempts to gain insight into electrostatic and magnetic phenomena. In Gilbert's

De Magnete is contained many experiments carefully described which could be repeated by later scientists to check and build upon Gilbert's findings. Although Peregrine limited the scope of his treatise to the properties of lodestone and terrestrial magnetism, his concept of the magnetic field was monumentally important, for it was through elaboration of this concept by later physicists that the behaviour of magnets became understood quantitatively, and this was vital to the development of the technology required in electricity generation for supply.

Gilbert's study was also important in making experiments in magnetism and electricity fashionable. Their bizarre effects must have been intriguing to those who had seen or heard of them and no doubt this inspired men like von Guericke to try their own experiments. Von Guericke experimented with electrically charged bodies a few decades after Gilbert. Guericke's work was a landmark in the sequence of events leading to the electromagnetic generators, because he was the first to succeed in devising a simple rotating electrical generator. Intent on facilitating the work of electrifying objects in his experiments, he designed a device which could accumulate electric charge using friction. The charge could be drawn off at will with a separate globe for experiments. Later, improved versions of the electric generator in the cylinder electric machine and the revolving doubler appeared which made electricity generation more efficient.

Electrostatic machines were important for two reasons. The first is that they were the earliest devices producing continuous electricity, though of course in the very nature of electrostatic electricity the quantities generated were very small. The current generated was of the order of 1 mA, with a voltage of 3×10^4 volt equivalent to 3×10^{-2} joule/sec. In this study a replica electrostatic machine- a van der Graaf generator- was made which showed the feasibility of such machines in producing a high enough voltage to produce discharges through the air.

The invention of the Leyden jar was important in leading to an understanding of the nature of electric charge. Even though a model made by the author had a very low capacitance by modern standards, as a device for storing electrostatic electricity the

Leyden jar was as important in the development of the electrostatic generator as the voltaic pile, storing current electricity, was in the development of electromagnetic generator, particularly in the pioneering work of Oersted and Faraday. The Leyden jar, acting as a kind of storage battery for electrostatic charge, amplified the effects of the electrostatic generator, but still the quantity of electricity supplied by this means was very small, as my model also indicated, and the science of electromagnetism could be not properly investigated until a means of providing more electricity could be discovered.

The work of Galvani and Volta on the production of electricity by chemical action, after considerable controversy over the source of the electricity causing muscular contractions in animals, led to the production of a battery of zinc and copper with separators soaked in an electrolyte, which produced a much larger continuous electric current than electrostatic machines. As shown by the results of the author's experiments with a replica voltaic pile, the voltaic pile could supply a comparatively strong current for a limited period, but was plagued by polarisation and local action troubles, which tended to reduce the performance of the battery to a low level in a short time. This necessitated the stripping down of the battery to laboriously clean the many plates and re-soak the separators in electrolyte. Nevertheless, this inconvenient source provided the tool with which the magnetic effect of a current was discovered, and later improved batteries made possible by Volta's work were an indispensable part of the experimental work of Sir Humphrey Davy's which led to the discovery of electric arc lighting before the work of Hans Christian Oersted. Like some other great scientific discoveries where the scientist is completely immersed in his scientific endeavour, Oersted discovered the magnetic effect of a current fortuitously.

Soon after, the first electromagnet was produced by Sturgeon, and it remained for Michael Faraday to demonstrate from the philosophy recorded in Oersted's research that phenomena in nature were interrelated and reversible could be applied in the case of magnetism and electricity. Faraday experimented with wires, magnets and galvanometers in a great variety of configurations and orientations which he was convinced would eventually lead to the discovery of what was later to become known as electromagnetic induction.

Not only did he make this discovery, but he showed how the principle could be applied in rotating machines to produce a continuous electric current, just as von Guericke had shown how to utilize the principle of electrostatic induction in rotating machines to produce electrostatic electricity. Faraday used his principle in two configurations, the rotating disc machine and the rotating loop. It seems incredible that at that time Faraday was able to detect the tiny current induced in a simple, single loop of wire rotating in the Earth's weak magnetic field, but with the astatic galvanometer at his disposal even such small currents as this *could* be detected, as was shown here. Faraday's disc generator was also of fundamental importance, because it was based on the principle used by modern generators- the use of a strong magnet whose lines of force were cut by moving conductors. The conductors comprised a solid disc in Faraday's machine, and its magnet was a permanent horse shoe magnet, which resulted in a higher e.m.f. than a simple rotating loop. Even though, as the author's model showed, the output current was impractically small, Faraday's machine pointed the way to the production of more powerful generators, suggesting the use of more powerful magnets and multi-turn coils, two innovations which were incorporated in Hipployte Pixii's generator which followed. In about 10 short years after Faraday's rotating disc machine, Hippolyte Pixii succeeded in making an magneto-electric generator which produced a much larger output. In this study, a model made of the generator produced 0.2 volt and 25 mA, whereas the Faraday model only managed 1 mV and less than 1mA. Besides the need to provide more powerful magnets and more turns of wire, a fundamental problem, which faced the electrical engineers of the first half of the 19th century, was how to dispose the coils in the magnetic field to maximum effect.

The answer was to wind the coils on a former, its ferromagnetic material increasing the density of the lines of force in the armature 100's of times, so the effect of the magnetic field was greatly enhanced. The complete arrangement was called the 'armature'. In the generator, the armature coils rotated in the most concentrated part of the magnetic field. Compare this arrangement with the disposition of the coils used in Pixii's machine mostly outside the magnetic field. The effectiveness of Pixii's machine was reduced because the coils for the most part were too far way from the poles of the magnet, and hence the field at the coils, which drops off with disance^2 was weak, lowering output voltage.

By the 1880's, the use of permanent magnets in generators was becoming unsatisfactory because of the limited strength of the magnetic field that *permanent* magnets of the time could provide, and this field strength deteriorated due to the demagnetising effect of the field produced by the induced current in the armature windings. Sturgeon's discovery provided the answer. Stronger magnets could be produced by using many turns of copper wire wound on soft iron cores, and the resulting electromagnet energized with batteries. Shortly afterwards, Varley designed a generator which energised its own electromagnet field magnets by diverting some of the current generated in the armature through the field windings: this was called self-excitation. From the 1850's development had reached a stage which attracted interest in the prospect of using electromagnetic generators as a source of electricity for arc lighting to replace impractical battery power.

An early attempt to provide light houses with arc lamps supplied by steam driven generators was successful and enterprising men like R.E.B.

Crompton determined to show that arc lighting supplied by generating machines could be made commercially viable, and that many were the advantages of arc lighting compared with the then almost universally established gas lighting. Crompton and Burgin also made significant improvements to the armature.

Whilst Crompton was lighting public entertainments, railway stations and private houses in this country, Thomas Edison in the States and Swan in the UK were independently developing the incandescent filament lamp which eventually enabled domestic buildings to be lit on a large scale. Edison also made important contributions to electricity supply with his bipolar dynamo and electric cable distribution systems, which were significant in solving some of the problems encountered in the generation and transmission of electric power. The most adventurous of his achievements was the opening of the first central generating station in New York, supplying street lighting and domestic lighting, which was important in showing doubting fellow electrical engineers throughout the world that electricity could be supplied successfully from a central generating station to consumers over a wide area. This was also the intention of the engineer who not only had the biggest ideas of all, but was successful in making them practicable: S.Z. de Ferranti.

S.Z. de Ferranti's first major contribution was the 'zigzag' armature alternator in which all available space in the generator was used up by densely packed electromagnets and an armature entirely made from copper. This machine was claimed to be ten times more efficient than other generators of the time. Ferranti's ambition was to build a central generating station outside London and transmit electrical power to consumers on a large scale all over London. However, since the 1870's when the commercial viability of electricity supply had been shown, numerous small electricity supply companies had sprung up to jump on the band wagon, and Ferranti's ideas were in direct conflict with *their* interests. When Ferranti had become a successful engineer with financiers backing him and he proposed his ideas for high voltage A.C. and centralized transmission to the commercial electricity-producing world, it was therefore not surprising that his big ideas came under attack. He was principally criticised for the suggestion that high voltage electricity could be distributed more efficiently that low voltage electricity. There were eminent advocates on both sides, and the *Battle of Systems* ensued. Only after much debate and a struggle in the courts, did Ferranti win the day and justly receive the support enabling him to build a generating station at Deptford to supply consumers in London via an unprecedented voltage of 10 kV.

Ferranti's high voltage distribution system from Deptford was the forerunner of today's National Grid system in which HV transmission voltages upwards of 132 kV are used. Ferranti designed his own cables for use in the Deptford distribution system, and the 10 kV cables were manufactured by the Norwich Wire Company which had been manufacturing cables since 1887. Cable technology around 1900 employed the basic techniques which had been used for manufacturing telegraph cables since the 1830's. However, by 1920 there was a need to increase voltage due to increasing demand, and this resulted in new problems in cable design which, with ever increasing voltages, have only more recently been solved.

Since Deptford in 1891, voltages have risen form 10 kV to 400 kV and maximum demand for electricity has increased form 10 MW to 70 000 MW (70 GW) in this country:

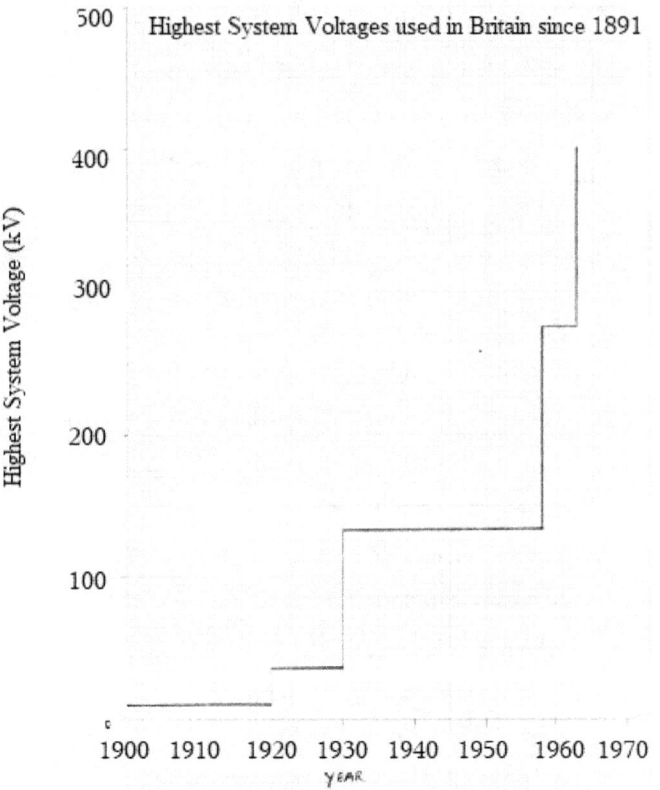

FIG. 21 GRAPH SHOWING HIGHEST SYSTEM VOLTAGES USED IN BRITAIN SINCE 1891

The growth in demand is exponential, doubling every ten years:

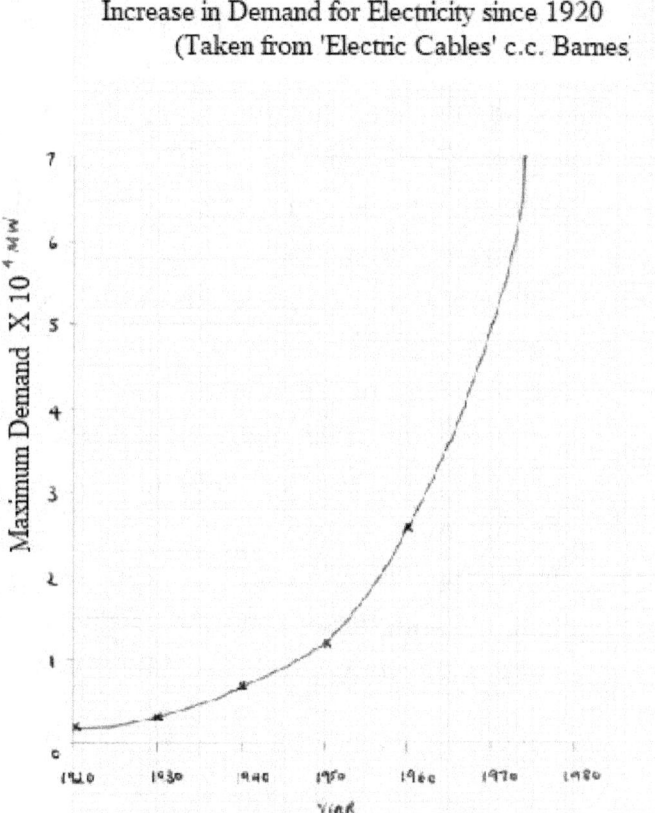

FIG. 22 GRAPH SHOWING INCREASE IN DEMAND FOR ELECTRICITY SINCE 1920

To cope with increasing demand, either very expensive conventional cable systems must be provided or some radical new approach must be made. Some scientists think that future demand could be met by developing superconducting cables. For example, a 100kV cable 1 cm in diameter at 4°K could carry 10 000 MW- enough to meet nearly 1/4 of the country's average consumption. Compare this with the rating of conventional 275 kV cables with a conductor area of 0.55 in^2 of 280 MW. But in view of the prohibitive expense and power losses involved in providing helium cooling to obtain the low temperatures required in these cables, new technology is focussing on High Voltage Direct Current transmission and the production of cables capable of carrying 7 GW (7 000 MW) at 800 kV.

10. BIBLIOGRAPHY

Abbot, A,Nelkon,M. (1971) 'Elementary Physics' Heinman

Admiralty 'Handbook of Wireless Telegraphy'

Barnes, C (1954) 'Electric Cables' Pitman London

Black, R.M (1972) .'Electric Cables in Victorian Times' HMSO

Bowers, B (1969) .'R.E.B. Crompton' Science Museum HMSO

Bulman, A. (1967) 'Model-Making for Young Physicists' London John Murray

"   (1972) 'Experiments and Models for Young Physicists' London John Murray

Cohen, B. (1956) 'Franklin and Newton' Harvard University Press

Derry, T., Williams,T. (1961) 'A Short History of Technology' Oxford University Press

_____'Rise of the Supply Industry' Geography Today

Grieg, D. (1969) 'Electrons in Metals and Semiconductors' McGraw-Hill

King, A., Wentworth, V. (1954) 'Raw Materials for Electrric Cables' Ernest Benn Ltd

Larsen, E. (1971) 'A History of Invention' Den & Sons London

Sootin, H. (19580 'Michael Faraday' Messner

Taylor, L. (1959) 'Physics, The Pioneering Science' Dover Publications

Tricker, R. (1965) 'Early Electrodynamics' Oxford Pergamon Press

Trigger, P. (1990) 'An Investigation of Two Electricity Topics in Teaching Undergraduate Engineering Science' Ed.D Dissertation, University of London.

_____(2013) 'Wind Turbines: Description, Appraisal & Alterantives. Vol I KDP Durham

Turner, D. (1933) 'The Book of Scientific Discovery' Harrap

Tyler,_ (1947) 'A Laboratory Manual of Physics' Edward Arnold & Co.

Wolf, A. , Dannerman, F., Armitage, A. (1935) 'A History of Science, Technology and Philosophy' Allen & Unwin

APPENDIX A
A REPLICA VAN DER GRAAF GENERATOR

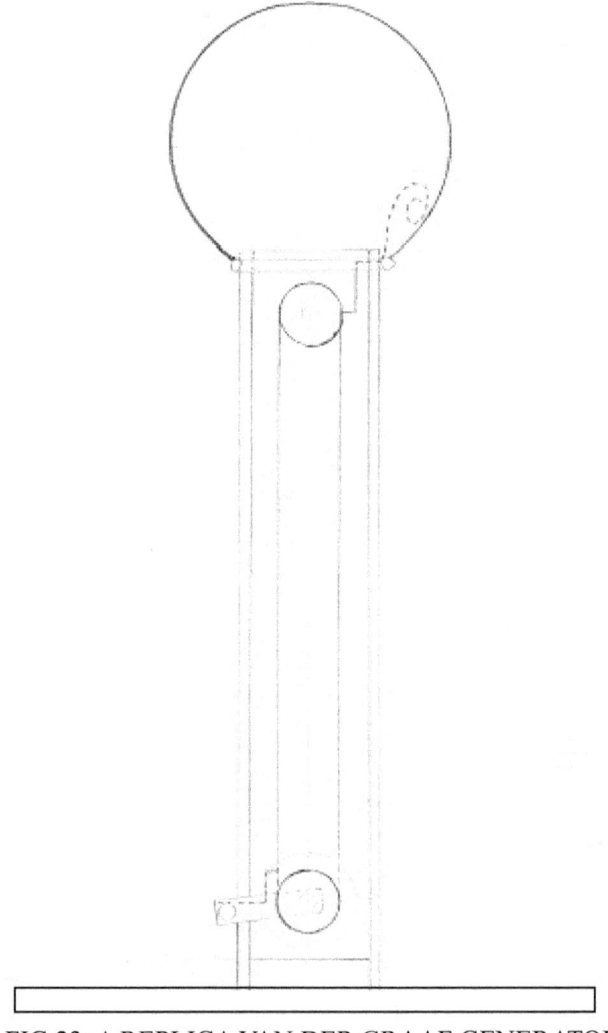

FIG 23: A REPLICA VAN DER GRAAF GENERATOR

A baseboard measuring 3/4" x 18" x 9" was used for the machine. (see above). Two pieces of perspex about 2 1/2" wide and 15" long were available and these were used to make the column of the machine, which was to contain the rollers and driving belt. Two blocks of the same material as the baseboard and screwed together provided an anchorage for the column. One block was cut 2 $\frac{1}{2}$" square and the lengths of perspex screwed to it so that they were separated a distance equal their width, forming a square-section column. The other block was made to be of the same width but was 1" longer and was screwed to first block and the baseboard. The column was supported at the top by a 1" wide perspex strip screwed to the drilled and tapped perspex column members.

The rollers were made from mild steel, 1" diameter cut to length so that there was sufficient clearance between the rollers and the sides of the column. Each roller was centre drilled on a lathe so that the 5/16" diameter mild steel rod used for the axles was a close fit in it. The perspex column was drilled at the top and bottom to allow the axles to turn freely in the column. The axles were equipped with adjustable stops made from 1/2" diameter alloy rod drilled and tapped so that a grub screw could be fitted. The bottom axle was fitted with a 2" *Meccano* pulley so that it could be belt-driven by an electric motor. To collect the charge at the top of the column a strip of fine brass wire mesh stripped at one end to produce a series of points was fixed under the perspex column support. The mesh was slotted so that it could be adjusted close to the belt. The other end of the mesh was wound into a spiral to enter the top conductor to make good contact with it. Another similar comb was fitted to the bottom of the column to spray charge onto the belt as it moved. The lower comb was connected to the positive terminal of a 1kV voltage source. The belt itself was made from a length of rubber tourniquet, sticking the ends together with rubber cement solution.

The top conducting sphere was made from a glass globe previously used as a light diffuser. Initially, it was coated with aluminium paint, but it was found that the conductivity of the globe was not good enough to produce sparks from the machine. A coating of 'Aquadag' colloidal graphite was applied and this produced a marked improvement in the conductivity of the globe.

The mouth of the globe was insulated from the air to restrict discharge from areas of small radius using split rubber tubing glued on to it.

PERFORMANCE

Determination of the Maximum Voltage Produced by a van der Graaf Generator

A 2 cm brass pendulum bob screwed into a mild steel rod 15 cm long was clamped into a retort stand and brought up to the globe of the van der Graaf generator. The maximum air gap which the sparks from the machine could be made to jump was measured to be 2 cm. by a wooden ruler placed behind the globe and the bob. Using Heydweiller's table (Table 1) below:

Radius of Balls	Distance Between Balls			
	0.1	0.5	1.0	1.5
Centimeter	Volts.	Volts.	Volts.	Volts.
2.5	4500	18900	33840	47610
1.0	4860	18020	32120	41160
0.5	4050	17790	17810	22400
0.25	4980	16200	20790	22980

TABLE 1 SHOWING THE RELATIONSHIP BETWEEN THE LENGTH OF SPARK IN AIR BETWEEN SPHERES OF VARIOUS AND DIAMETERS AND THE VOLATGE BETWEEN THEM AT 15 DEG C AND 76 CM OF MERCURY PRESSURE

Source: Heydweiller

this corresponds to a voltage of at least 40,000 volt.

APPENDIX B

A REPLICA LEYDEN JAR

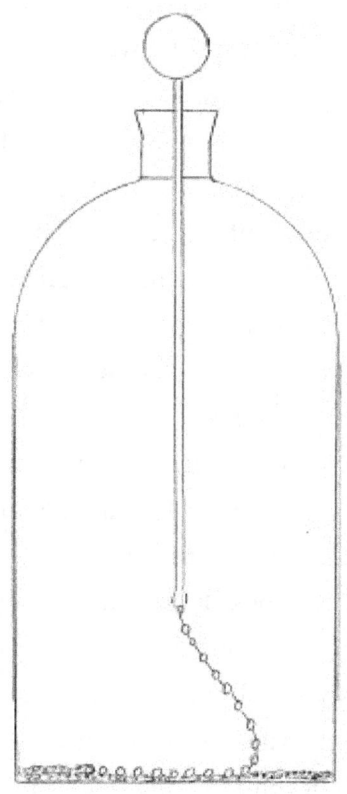

FIG 24. DIAGRAM OF THE MODEL LEYDEN JAR

CONSTRUCTION

Because of the difficulty of introducing a metal foil electrode into as glass bottle, to offset the effect of poorer conductivity of the inner electrode used (see below) compared with that of the early Leyden jar, a thin polythene container was used instead (see above). This was thoroughly cleaned and dried. Somehow a conducting medium had to be arranged on the inner surface of the container to act as one electrode, and any such operation had to be carried out with mercury. As sufficient quantity of mercury was not available and the mercury vapour intrusive, this was decided against. However, a layer of black lead would provide a conducting surface, and the former was obtained in the form of 'Zebrite' grate polish in paste form. To stick the coating to the inside of the container, glue was mixed with the paste, which also made it less viscous and easier to apply. But the mixture dried very quickly with the result that, applying it with a paint brush produced an uneven layer. The final substance used most successfully was colloidal graphite 'Aquadag' (as with the replica van der Graaf machine), which could be diluted with water and applied like paint. This resulted in a complete and uniform layer on the inside surface of the 'jar' and it provided improved conductivity compared with the black lead/glue mixture which was measured as 2 x $10^{-5}\Omega^{-1}$cm. One ounce of glued lead shot was introduced through the neck of the container and fixed to the bottom to assist conductivity and stabilise the 'jar'. The centre electrode was to be free from sharp points outside the jar, and the top conductor therefore needed ideally to be a sphere of about 2 cm diameter and not so heavy as to make the lightweight jar unstable to toppling over. After considerable searching, an old brass pendulum bob equipped with a hook was found. The hook was sawn off flush with the sphere and filed to complement the shape of the sphere. A rod of about 3/16" mild steel the jar its lower end was about 2" from the bottom and its top protruded 2". A hole 0.005" larger than the diameter of the rod was drilled in the brass sphere to a depth equal to its radius. The periphery of the sphere near the hole and the end of the rod were cleaned with emery cloth. The tip of the rod was re-heated and a layer of solder applied. The tip of the rod was heated and the steel rod pushed home in the hole. Excess solder was turned off on the lathe and to clean up the surface of the rod and sphere. A 3/32" hole was drilled in the other end of the rod, and the hooks of a sink plug chain pressed through the hole. The chain was to rest on the bottom of the jar to improve the central conductor's effective area of contact with the graphite coating inside the jar. The central conductor was held in position by a rubber stopper pressed into the neck of the container. The outer conductor was made by wrapping a strip of aluminium cooking foil onto the container, covering the entire flat area of the outside. All edges were insulated with sticky tape.

PERFORMANCE
Determination of the Capacitance of the Leyden Jar

The circuit used is shown below:

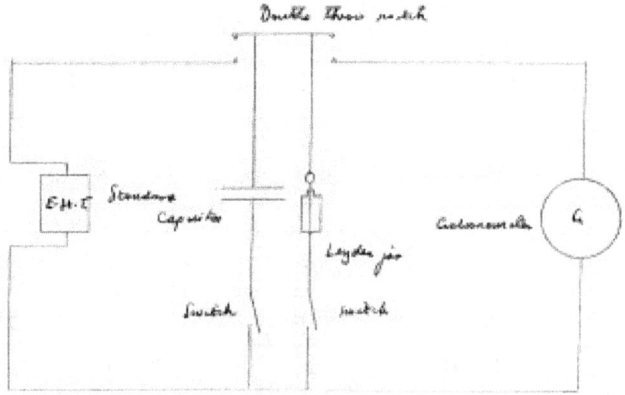

FIG 25. CIRCUIT USED TO DETERMINE THE CAPACITANCE OF A LEYDEN JAR

At first a 1.5 volt cell was used to charge the capacitors, but the small capacitance of the Leyden jar and the resistivity of the coatings used for its 'plates' did not hold a measurable charge. A variable E.H.T. source of 0.6 kV was used, and it was found that after charging the Leyden jar to 1 kV a half full-scale deflection on the galvanometer was achieved. Various standard capacitors were tried in this manner keeping the voltage at 1 kV and it was found that a 0.0001 µF gave a comparable deflection to the Leyden jar. The maximum deflection caused by the fully charged standard capacitor and Leyden jar (θ) was determined and the capacitance of the Leyden jar calculated.

Results

The galvanometer deflection using a standard 0.0001 µ F capacitor: $\theta = 22$; Leyden jar: $\theta = 20$

Calculation

Since the maximum throw of the galvanometer is proportional to charge, the calculated capacitance of the Leyden jar is:

0.0001 µF x 20/22 = 9 x 10^{-5} µF

Determination of the Maximum Voltage to which the Leyden Jar could be charged

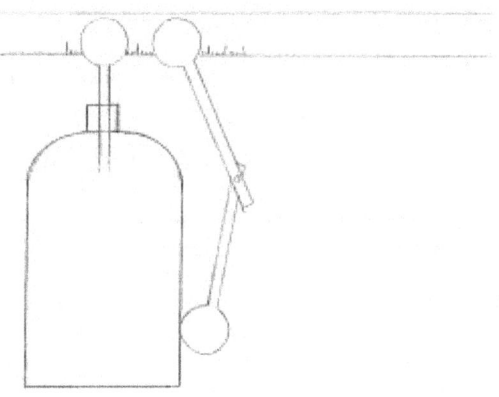

FIG 26. DETERINATION OF MAXIMUM VOLTAGE OF A LEYDEN JAR

The experimental setup is shown above. Table 1 shows the relationship between the length of spark in air between spheres of various diameters and the voltage between them at 15 deg C and 76 cm of mercury pressure. No account of humidity is taken in the table, but the compiler, Heydweiller, explains that the voltages in the table must be increased by 1% for a fall of 3°C and 1% for a rise of 8mm pressure. From this it appears that, setting relative humidity aside, which would be low anyway in a centrally heated room in which the experiment was carried out, variations in temperature and pressure have little effect.

Results

The relevant voltages in the table are 33 840 volt and 32 120 volt since the spark gap jumped in the experiment with the Leyden jar was 1 cm, and the radius of the spheres 2 cm, i.e., between 2.5 and 1 cm.

Calculation

The maximum voltage of the Leyden jar is therefore somewhere between 33 840 V and 32 120 V, and with a measured temperature of 15°C and atmospheric pressure of 74 cm, the error in the voltage in the table is about 5%. Since the difference between the relevant voltages is 1720, and 5% of this is less than 100 V, allowing for error, the maximum voltage of the Leyden jar still lies between 32 120 V and 33 839 V.

Determination of the Maximum Charge of the Leyden Jar

To find out how much charge the Leyden jar could hold, and what voltage it could be charged to, the circuit in Fig. 25 was used. The capacitance was previously found to be 9×10^{-5} µF and the maximum voltage about 30 kV. The maximum charge of the jar was calculated as follows. The value of CR, the time constant of a discharge circuit containing a capacitance C and resistance R is the time taken for the voltage of the capacitor to fall by 1/e or about 1/3. The Leyden jar when discharged via a length of copper wire of resistance $0.01 \; \Omega$ connected between its terminals discharges in approximately $9 \times 10^{-5} \; \mu F \times 0.01 \; \Omega \approx 10^{-12}$ sec. The charge on the jar is therefore:

$Q = CV = 9 \times 10^{-5} \mu F \times 10^{-6} \times 3 \times 10^{4}$ coulomb $\approx 2.7 \times 10^{-6}$ coulomb.

APPENDIX C

THE ORIGIN OF ELECTRICAL POTENTIAL

When a metal is immersed in water, there is a tendency for metal atoms to form ions which go into solution, and electrons which remain on the metal. Metal ions in solution will in like manner be deposited on the metal tending to decrease its charge:

$M \to M^{n+} + ne^-$
(1)
Metal   ions in solution

$M^{n+} + ne^- \to M$ deposited on the metal
(2)

i.e. $M \leftrightarrow M^{n+} + ne^-$.

When the rates of reaction of (1) and (2) are equal, equilibrium results and a final stable value of P.D. between the metal and the solution results. The position of the equilibrium (how far the reaction goes towards the right) depends on the ionization energy of the metal: its readiness to lose an electrons to form positive ions. Thus highly metallic elements in the top left hand corner of the Periodic Table of elements such as the alkali metals would push the equilibrium in (2) far to the right. In this case, there is a greater tendency for ions to go into solution than to be deposited back on the metal, and so the metal will be negatively charged with respect to the solution. To standardise this P.D., it is assumed that the potential of the metal electrode is at earth potential (see Fig. 27). Metals such as copper, which is far less reactive than the alkali metals, have less tendency to form positive ions, so that for copper the equilibrium in (2) is towards the left. The result is that the solution becomes negatively charged with respect to the metal.

Besides the reactivity of the metal, the direction and size of the P.D. between a metal and solution will depend on the concentration of metal ions in solution. If two different metals are placed together in solution, electrons will flow from the metal having excess electrons, the negative plate, to the other metal, the positive plate. The P.D. between them will be their difference of potential with respect to earth in their separate solutions

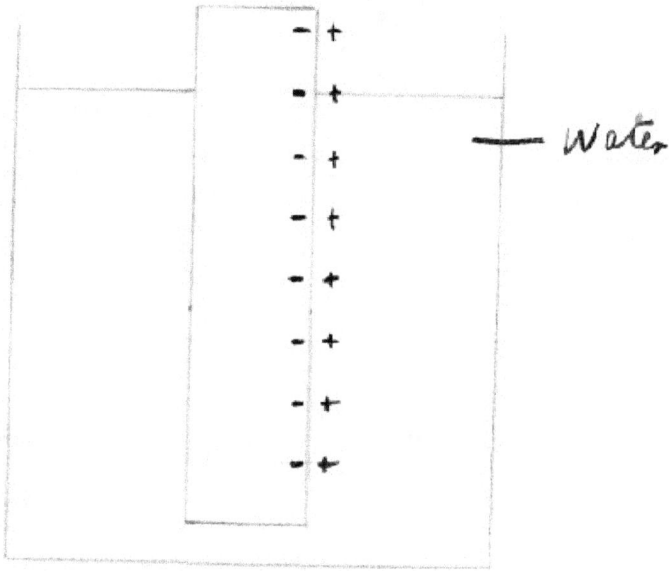

FIG. 27 IONIZATION OF A METAL IN WATER AND THE PRODUCTION OF A POTENTIAL DIFFERENCE

APPENDIX D

A REPLICA VOLTAIC PILE

FIG. 28 A REPLICA VOLTAIC PILE

DESCRIPTION

To make sure that enough current would be produced by the battery I used 19 pairs of plates of copper and zinc (see above). Both types of plate were cut from sheet metal and were 2" square. To keep the battery vertical a made a box-section former from 3/8" plywood about 12" long to hold the plates and insulators. The inside was varnished to prevent seepage of the electrolyte into the wood which would otherwise have shortened the life of the battery and ruined the former. The two ends, the electrodes of the battery, were soldered to wires that carried the current produced to the testing apparatus used. Three different materials were tried to hold the electrolyte: filter paper, blotting paper and thick towelling. The filter paper was not absorbent enough and caused the electrolyte to quickly dry out . The blotting paper tended to break up with the action of the corrosive electrolytes used. Volta must have experienced this problem, though perhaps it was less pronounced since he used cardboard. The thick towelling proved the most satisfactory since being thicker it absorbed more electrolyte. It also held the plates further apart so that they would not touch under the compression applied to the column to improve contact area between the plates and the electrolyte. A disadvantage was that the thicker material increased the battery's internal resistance; it also made it physically larger.

As Volta initially used brine as an electrolyte, brine was initially used here. It proved unsatisfactory because as the water in the brine evaporated it deposited salt on the electrodes causing first a dramatic reduction in the P.D. obtained from the battery and then a complete breakdown. Hydrochloric acid was found to be more satisfactory, and the thick paper towelling separators were soaked in a 0.2N solution of it.

PERFORMANCE

The circuit used and method of calculation used to determine internal resistance is shown on pp. 254-6. Firstly, Tables 2 and 3 show the results obtained for internal resistance.

Calculation of internal resistance for filter paper separator

L_1, cm	L_2, cm	R Ω	$r = R(L_1-L_2)/L_2$
67.2			
	17.2	30	87.2
	12.2	25	112.7
	9.2	20	126.2
	6.1	15	150.0
	2.4	10	

TABLE 2 CALCULATION OF INTERNAL RESISTANCE: FILTER PAPER SEPARATOR

Calculation of internal resistance for towelling paper separator

L_1, cm	L_2, cm	R Ω	$r = R(L_1-L_2)/L_2$
68.5			
	20.4	30	50.8
	19.4	25	55.7
	18.1	20	63.8
	15.6	15	70.6
	14.1	10	
	11.1	5	

TABLE 3 CALCULATION OF INTERNAL RESISTANCE: PAPER TOWELLING SEPARATOR

The lowest figure was obtained with the paper towelling separator. An average taken at the various loads was 60 Ohms.

Fig. 29 shows the circuit diagram of the circuit used to variation in voltage with compressive force on the plates with no load battery and discharge performance under load .

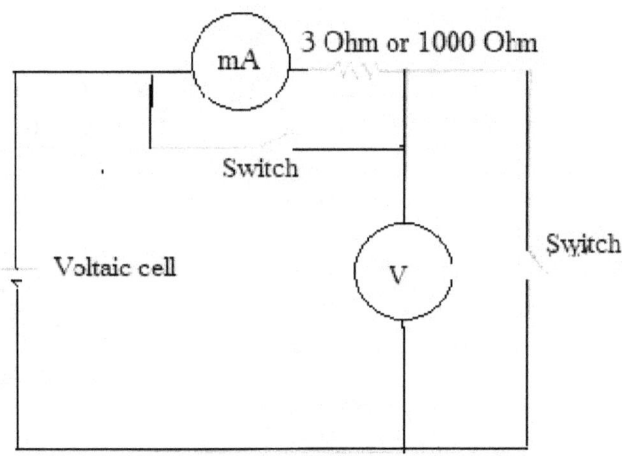

FIG. 29 THE CIRCUIT USED TO MEASURE BATTERY PERFORMANCE

Maximum Potential Difference

The variation of P.D. with compressive force on the plates by weights placed on top of the pile is shown in the table below:

VOLTAIC PILE

Potential Difference (volts)	Compression Weight (grams)
0.8	0
0.81	100
0.82	200
0.82	400
0.80	500
0.83	600
0.83	700
0.85	900
0.84	1000
0.85	1200
0.84	1400
0.84	1600
0.84	1800

TABLE 4 THE VARIATION IN P.D. OF THE VOLTAIC PILE WITH COMPRESSIVE FORCE ON THE PLATES

The relationship between compressive force and P.D. is graphed below:

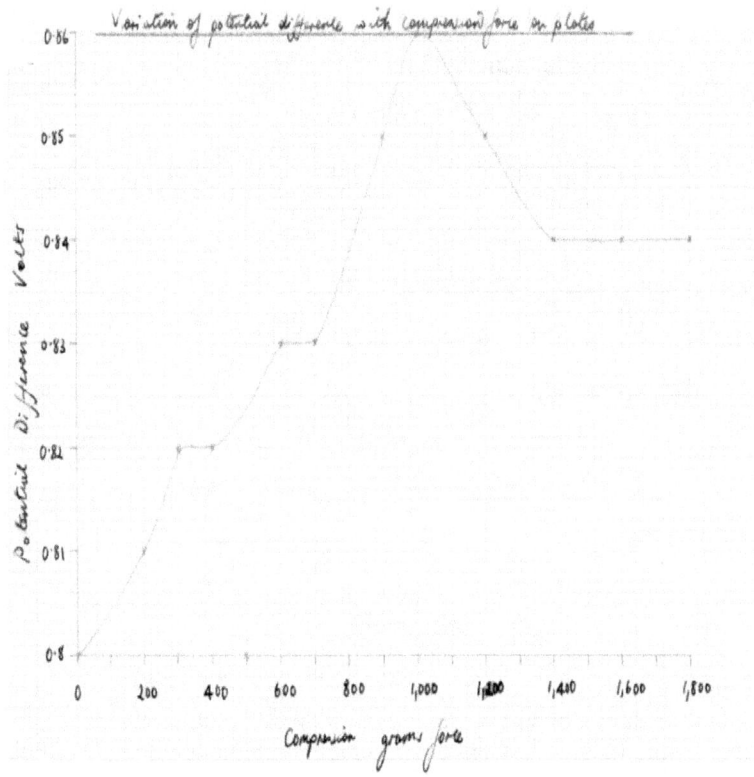

FIG 30 GRAPH OF VARIATION IN THE P.D. OF A VOLTAIC PILE WITH THE COMPRESSIVE FORCE ON THE PLATES

The maximum P.D. obtained was 0.9 volt. With an average lowest internal resistance of 60 Ω this meant that, using Ohm's law, a maximum current of

0.9/60 x 1 000 = 15 mA was obtainable from the battery. I tried to improve this value by using thinner separators of filter paper, but Table 2 shows that an average value of internal resistance of 120 Ω was obtained. The unexpected increase appears to have resulted from the poor absorptive capacity of the filter paper which was unable to maintain a film of electrolyte on the plates.

Variation of P.D. and Current over Time with Load

The Table below shows the performance of one voltaic cell with paper towelling separators into 3Ω:

TABLE 5 PERFORMANCE OF A VOLTAIC PILE (3Ω LOAD) : ONE CELL

PAPER TOWELLING SEPARATOR
PERFORMANCE; ONE CELL

Time of discharge	Potential Difference	Current	Load
0.00 minutes hrs	0.8 volt	23 mA	3 Ω
0.05			
0.30		20	
1.00	0.1	17.5	
1.45		16	
3.02	0.08	15	
10.02		14	
20.20	0.07	13	
28.20	0.06	11	
33.05	0.05	9	
43.00	0.02	4	
40.40	0.03	5	
38.45		6	
37.05	0.04	7	
35.12		8	

249

The data in Table 5 are graphed below:

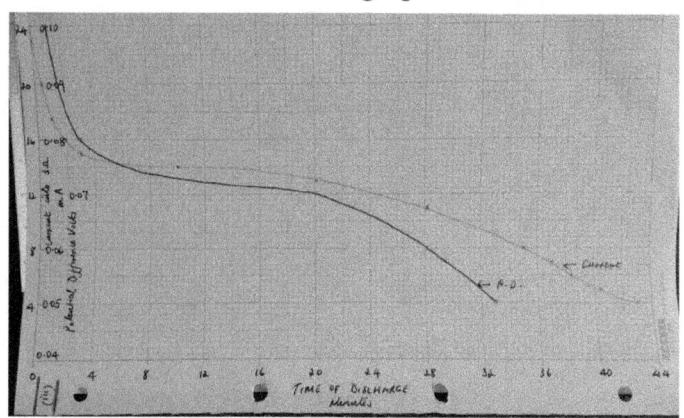

FIG. 31 PERFORMANCE OF A VOLTAIC PILE (3Ω LOAD) : ONE CELL (PAPER TOWELLING SEPARATOR)

(iv) Filter Paper Separator; Current into 3Ω			
Time of Discharge (hours)	Current (mA)	Potential Difference (volt)	Load
0	0	0.74	3Ω
0.10	17	0.1	
0.30	14	0.07	
0.58	11	0.06	
1.30	9	0.05	
2.08	8	0.04	
2.45	6	0.03	
3.00	5		
4.00	4	0.02	

TABLE 6 PERFORMANCE OF ONE VOLTAIC CELL WITH FILTER PAPER SEPARATOR INTO 3Ω

250

The graph below shows the performance of one voltaic cell with filter paper separators into 3Ω:

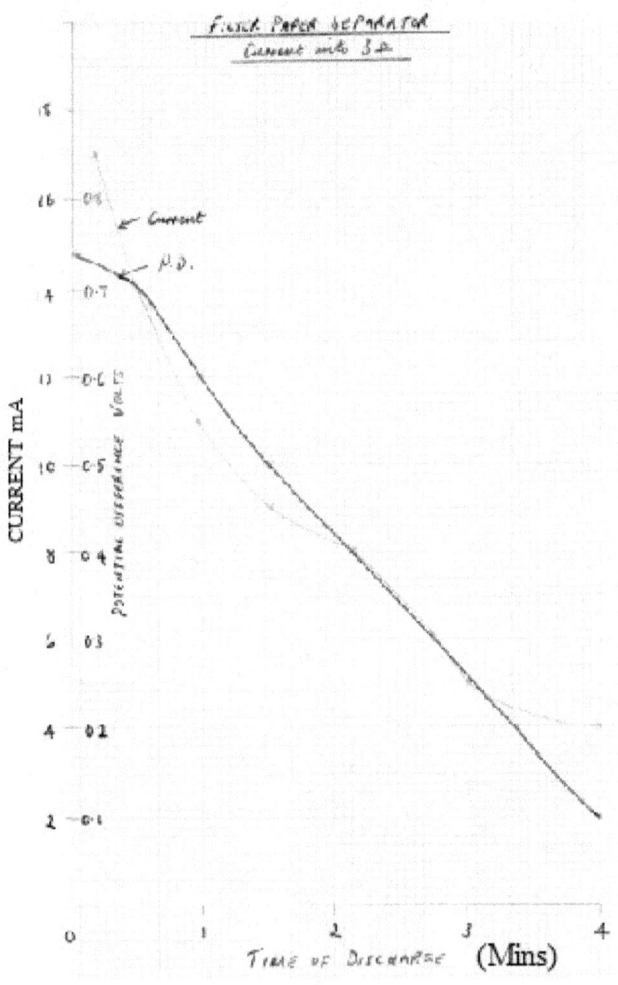

FIG. 32 VOLTAIC PILE: GRAPH OF PERFORMANCE WITH A 3 OHM LOAD

The table below shows performance into a 1,000 Ω load:

Time of Discharge (hours and mins)	Potential Difference	Current (mA)	Load (Ω)
0		0.64	1000
1.30		0.63	
3.05		0.62	
5.50		0.60	
6.25		0.59	
6.55		0.58	
7.25		0.57	
7.55		0.56	
8.25		0.54	
9.15		0.51	
9.45		0.50	
10.00		0.49	
11.07		0.47	
12.53		0.46	
16.00		0.45	
20.40		0.44	

TABLE 7 VOLTAIC PILE PERFORMANCE WITH 1 000 Ω LOAD

The data in Table 7 are graphed below:

FIG. 33 GRAPH OF VOLTAIC PILE: CURRENT VS TIME OF DISCHARGE

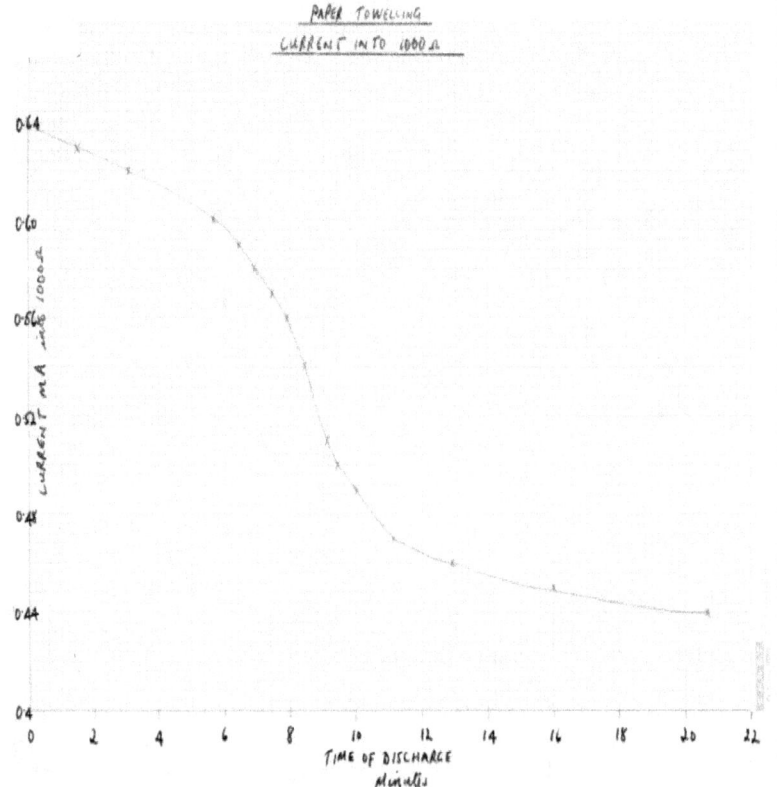

The 3Ω load was calculated to be the resistance representing the load on the cell equivalent to a typical load on a modern Leclanche type cell by a torch bulb. This was done to express the performance of the voltaic cell in modern day terms in comparison with modern cells. From Table 5 it can be seen that with the 3Ω load after 1 minute the potential difference drops to 0.1 volt. Also the current falls quickly during the first few minutes steadying at an output of 0.08 volt and 14 mA for 15 minutes.

Table 7 and Fig. 33 show the same variables using paper towelling separators and a 1,000Ω load. A fairly stable P.D. of around 0.6 V (using Ohm's Law) and 0.6 mA V in the first few minutes on discharge was obtained, reducing to 0.45 V and 0.45 mA respectively within 15 minutes. The very poor performance found using filter paper separators seems to have been mainly due to drying out of the filter paper. But with both types of separator, when the cell is delivering current, hydrogen gas produced by electrolysis of the hydrochloric acid collects on the zinc plates, forming a high resistance layer and so increasing the cell's internal resistance (polarisation). In addition, impurities in the zinc, such as iron, set up a cell of local action at the zinc plate resulting in electrolysis even when the cell is delivering no current. As a result, the previously bright surface of the zinc plate became insulated from the rest of the cell by a black coating and useful e.m.f. could no longer be drawn from the cell. To make the cell reusable, the 20 pairs of plates had to be scrubbed clean and the separators replaced and again acidified. After using the battery and putting it into storage the plates would have to be once again removed and cleaned: a laborious process for Volta as he coupled many batteries together in order to produce the high voltages required to charge Leyden jars, etc. In addition, the zinc plates become pitted with use, reducing the area of contact with the rest of the cell and increasing its already large internal resistance.

Comparison between the Performance of the Voltaic Cell with a Modern Leclanche Cell

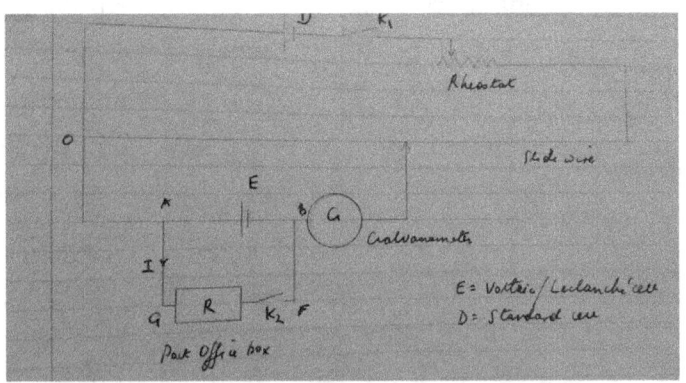

FIG. 34 CIRCUIT USED TO COMPARE THE PERFORMANCE OF A VOLTAIC PILE WITH A MODERN LECLANCHE CELL

The circuit diagram, using a Wheatstone Bridge arrangement, is shown above. k_1 was inserted and with k_2 open a balance point about 2/3 the way along the slide wire obtained. The balance point at c_1 distant l_1 cm from O was measured with a metre rule. 30Ω was taken out of the Post Office box and k_2 closed and a new balance point c_2 distance l_2 cm from O obtained. More values of l_2 were obtained by decreasing R by 5W each time. The internal resistance r was determined as follows.

At balance point c_1 $E \propto l_1$, where E is the e.m.f. of the cell.
" " c_2 $V \propto l_2$, where V is the P.D. between A and B.

Let the current in ABFG be I amp, then
$V = I.R = E.R./(R+r) \propto l_2$,

but $E \propto l_1$
Therefore, $(R+r)/R = l_1 / l_2$ from which

$$r = R \cdot \left[\frac{I_1 - I_2}{I_2} \right] \ \Omega$$

In the table below and the graph on p. 257 is shown the performance of a modern Leclanche type dry cell (a) internal resistance and (b) into 3Ω.

(a) Internal resistance, r:

I_1 cm	I_2 cm	R Ω	$r = R(I_1 - I_2)/I_2$ Ω
62.4			
	61.4	30	0.488
	61.2	25	0.448
	59.3	4	0.209
	60.9	15	0.370
	61.8	40	0.388
	59.6	5	
	58.6	3	

(b) Discharge characteristic (3 Ohm load):

Time hrs/min/sec	Current in A	Voltage (volts)	
0	.295	2.95	
26.30	.293	2.90	
56.05	.290	2.85	
1.44.00	.287	2.80	
2.39.00	.285	2.75	

TABLE 8 LECLANCHE CELL PERFORMANCE

It can be seen that its e.m.f. was steady: within -0.2 volt over the period of the test of nearly 3 hours, in contrast to the voltaic cell's regulation of -0.65 volt at best after just 4 minutes. Compared with modern day cells, the voltaic pile was crude, inconvenient and had a vastly inferior internal resistance and

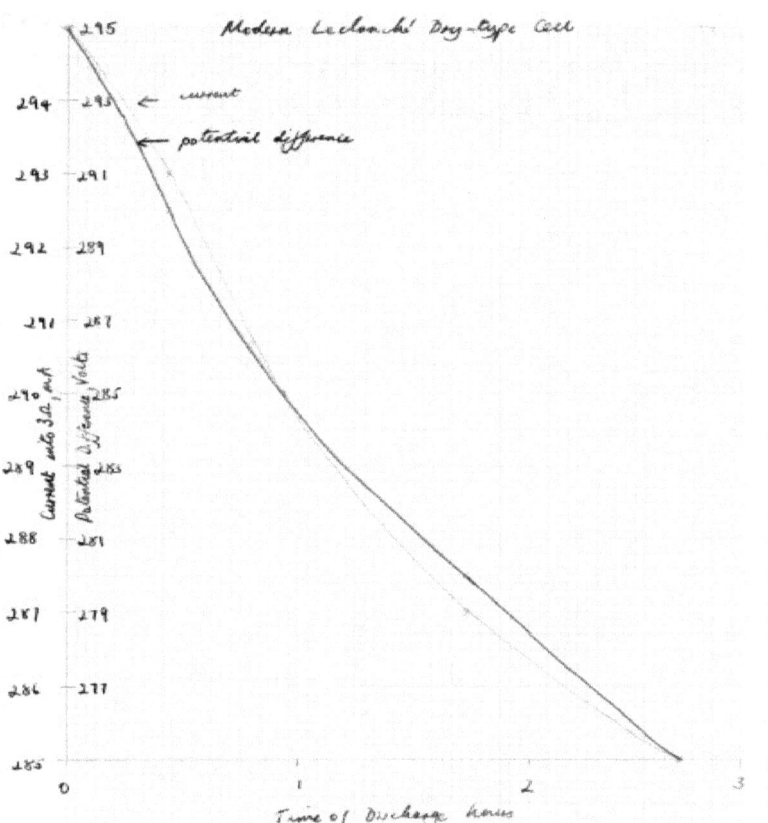

FIG 35 GRAPH OF CURRENT/P.D. OF A LECLANCHE CELL VS DISCHARGE TIME

ampere-hour capacity as shown in Table 5 in comparison with that in Table 8. This was primarily due to polarisation. Volta was aware of this shortcoming and experimented with other metals for the plates. Nevertheless, when twenty pairs of zinc-copper plates are connected together about 18 volt is produced into a high resistance load such as moist skin or a Leyden jar- a voltage easily detectable by these means.

APPENDIX E

FARADAY'S DISC GENERATOR

A REPLICA FARADAY DISC GENERATOR

The original machine was simple in construction., so the model's fabrication was straight forward and the details are shown below:

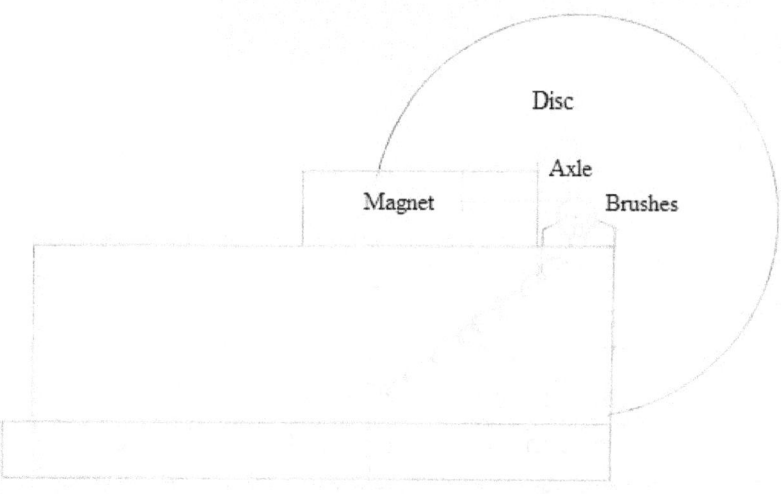

FIG. 36 A REPLICA FARADAY DISC GENERATOR: SIDE VIEW

FIG. 37 A REPLICA FARADAY DISC

GENERATOR: TOP VIEW

It was connected to an adjustable speed motor (though of course Faraday's machine had a handle) and run up to a speed at which the e.m.f. output was steady. The circuit used to measure the performance of the disc generator is shown below:

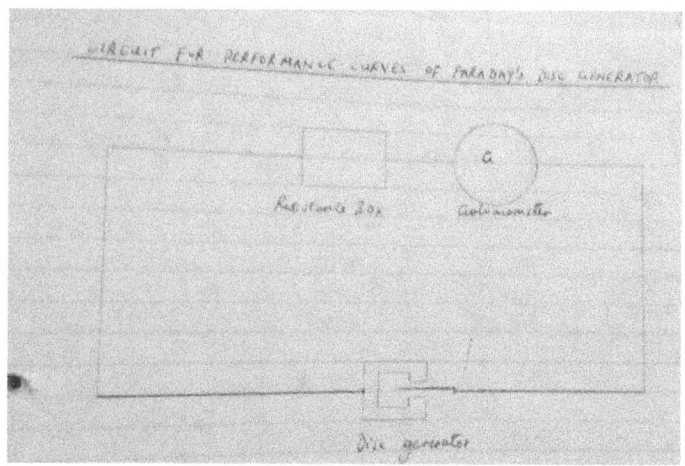

FIG. 38 CIRCUIT FOR PERFORMANCE TESTING OF A FARADAY DISC GENERATOR

A resistance box and galvanometer were connected in series with the generator so that variation in current with load could be determined. Without any load in circuit, the galvanometer reading was taken, and then the load increased in steps of 5Ω to 30Ω, then steps of 10Ω up to 80Ω and then steps of 40Ω up to a maximum of 120Ω. Performance graphs of current v load and current v 1/load were plotted, and e.m.f. was calculated from the slope of the latter at the origin. The tabulated data are shown in the table and the graphs following:

PERFORMANCE OF DISC GENERATOR

LOAD (Ω)	1/LOAD	CURRENT (μA)
Galvanometer only		454
5	0.200	396
10	0.100	330
15	0.067	308
20	0.050	264
25	0.040	242
30	0.033	220
40	0.025	176
50	0.020	154
60	0.0167	132
70	0.0143	110
80	0.0125	88
120	0.0083	66

TABLE 9 FARADAY DISC GENERATOR PERFORMANCE

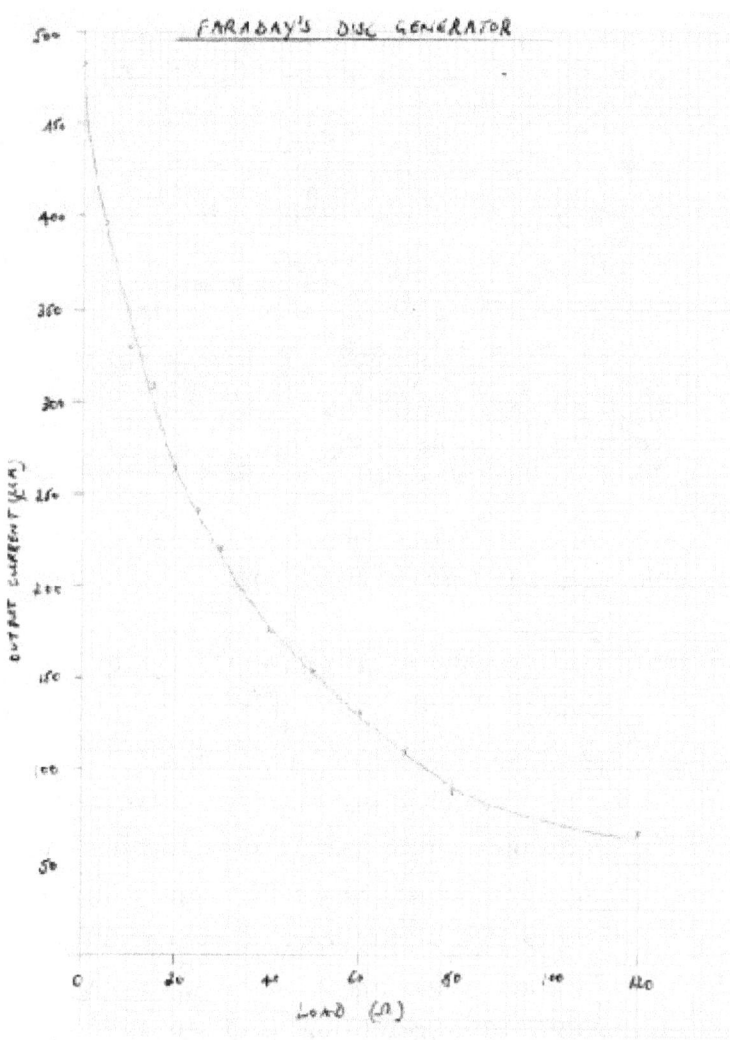

FIG. 39 FARADAY DISC GENERATOR GRAPH OF OUTPUT CURRENT VS LOAD

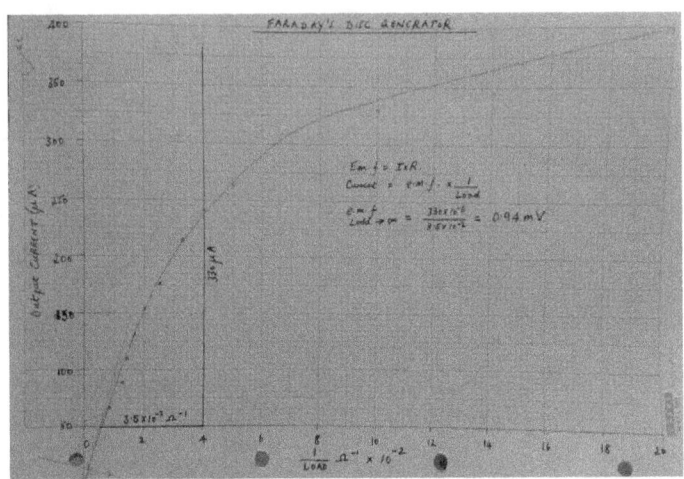

FIG. 40 FARADAY DISC GENERATOR GRAPH OF OUTPUT CURRENT VS 1/LOAD

The generator at first gave a low output of one or two mA and the reasons for this were twofold. The detecting instrument used initially was an 'Avometer' switched to the 50mA range full scale deflection. Unfortunately, the Avometer had a high resistance which protects it against high currents. Before this was realized, it was thought that the generator itself had a high resistance, but when measured without the machine running it was measured as a fraction of 1Ω. In fact, in operation the machine's resistance was higher than this with spring loaded brushes used as terminals. These did not achieve good contact with the machine since the disc was running slightly eccentrically. Added to this, even after trying springs of various elasticities for the brushes, the pressure required to achieve good contact made the machine run eccentrically. When the brushes were later exchanged for strips of spring steel, the characteristic of the machine improved considerably. A second reason for low output current might be too low a magnetic field strength from the magnet. Considering this:

the flux of the magnet was 167 000 Am^{-1},
the area of the gap $\qquad$ 4 cm^2.

Assuming a uniform field between the poles, the e.m.f. induced is given by:
$e = -d[BA]/dt$ $\qquad$ (Faraday's law),

where A is the area of the gap.

Hence $BA = \mu_0 H = 4\pi.10^{-7}.4.10^{-4}.1.67.10^5$ Wbm^{-2}
$= 8.4.10^{-5}$ Wbm^{-2}.

With the magnet placed with its poles at the rim of the disc, a sector of arc length 3 cm cuts the field completely.

Therefore, if the circumference of the disc is 10π cm, the field is cut completely $10\pi/3$ times per revolution.

Hence, if there are ω revs or the disc per second then

$\varepsilon = 10.8.4.10^{-5} . \omega$ volt.

With the resistance of the machine measuring about 1Ω, the current on open circuit should therefore be about 840 µA @ 60 r.p.m..

On open circuit and with the machine speed adjusted to give a steady output, it produced about 600 µA when connected directly to the galvanometer. Owing to eccentricity in the disc, at speeds below optimum, the pressure of the spring steel pick-up terminal varied with the position of the disc in its revolution and the speed of the machine, and hence its output fluctuated. At higher speeds, more e.m.f. was produced, but vibration of the axle of the disc in its bearings and the inefficient contact of the terminal at the rim (again due to eccentricity in the disc), did not produce an e.m.f. which rose in proportion to the rate at which magnetic flux was cut by the disc, .i.e., the speed of revolution, and the highest e.m.f. that could be obtained at any practicable speed was about 1 mV.

APPENDIX F

GALVANOMETER SENSITIVITY

If we assume the galvanometer available to Faraday at the time had a former of 3 cm radius and 1 000 turns, the magnetic field at the centre using the Biot-Savart law will be:

$$\Sigma \mu_0 \cdot dl \cdot i \cdot \sin\theta / r^2,$$

where μ_0 = permittivity of free space,

i is the current in the coil, dl the length of a current element and a line connecting the element with the centre of the coil.

For a circular coil, $\theta = \pi/2$, and evaluating the above sum S for each element gives the expression:

$$B = 2.\pi.\mu_0.i.N/r,$$

where N is the number of turns.

Numerically,
$B = 4.\pi.10^{-7}.2.\pi.i.10^{3}.10^{-2}/3 \ A^{-1}Wbm^{-2}$
$\approx 2.5.10^{-8} \ mA^{-1}Wbm^{-2}$ (1).

If the coil is set in the magnetic meridian a balance between the Earth's magnetic field and the coil's magnetic field can be obtained, resulting in a steady deflection of the galvanometer needle.

The Earth's field is $\approx 0.2 \cdot 10^{-7} \cdot 4\pi$ Wbm^{-2}
If the galvanometer is allowed to come to equilibrium,

$H_0 \cdot \tan\theta = B$ (coil) (2),

where H_0 is the Earth's field strength and θ is the angle of the needle to the meridian.

For a deflection of 1°,

$H_0 \cdot \tan\theta \approx 0.2 \cdot 4\pi \cdot 10^{-7} \cdot 2 \cdot 10^{-2} \approx 5 \cdot 10^{-9}$ Wbm^{-2}

Therefore, from (2),

$B \approx 5 \cdot 10^{-9}$ Wbm^{-2} for a deflection of 1°. (3).

Comparing (1) and (3) the magnetic field produced in the galvanometer coil carrying a current of a mere fraction of 1mA deflects the galvanometer needle perceptibly. Thus with an instrument of this sensitivity, Faraday could have detected the feeble currents produced by electromagnetic induction in a single loop in the Earth's field.

APPENDIX G

A MODEL REPLICA PIXII GENERATOR

Pixii's original generator used a permanent horse shoe magnet to induce an e.m.f., and this was revolved across the pole faces of two coils. The magnet available for my home-made machine was attached to a driving spindle which I thought could be accomplished by drilling the magnet and inserting a metal rod, using epoxy resin to make a secure joint:

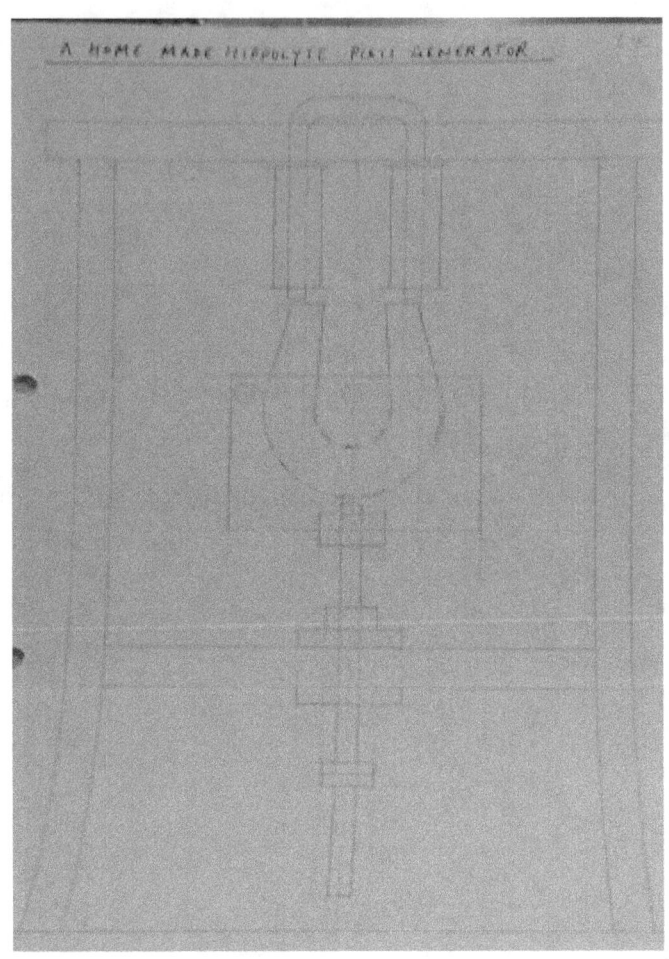

FIG . 41 A REPLICA PIXII GENERATOR

However, I found the material of the magnet too hard too drill, and so to hold the magnet a special saddle was made up using a square *Meccano* metal sheet bent double, and three long screws and nuts. The magnet was placed part way down in the hollows of the saddle and clamped in the desired position by tightening the screws and nuts. The 6 mm diameter spindle was fixed in the bottom of the saddle at right angles to it by means of two adjustable stops. This arrangement had the advantage that the pole pieces of the magnet could be adjusted to be accurately parallel to the coil pole faces. This enabled the use of a minimal working clearance and therefore a small air gap between the two sets of pole faces, resulting in a stronger magnetic field density in the armature coils. However, in Pixii's original generator, there was no flexible connection between the driving spindle and the magnet.

The coils and magnet were supported vertically so that a cage-like structure had to be made to hold the coils above the magnet, and to support a bearing in which driving spindle was to run. This was accomplished by cutting a baseboard of dimensions 30 x 20 cm and arranging four metal rods vertically at the corners of a 10 cm square.

Fortunately, some old retort stand uprights were available whose ends carried a 1/2" B.S.W. thread. These were cut to 13 cm lengths and the base board drilled and tapped to accommodate them. The bearing plate was made from the same material as the base board and cut to fit the cage snugly. The bearing was made by drilling a 1/2" hole in the bearing plate and fixing two rectangular strips of perspex on to the surfaces of the bearing plate above and below the hole. The Perspex bearing was drilled to accommodate the driving spindle. To fix the bearing plate in the cage, the cage uprights were drilled 5/16" and the bearing plate secured by wood screws. By means of this arrangement, the orientation of the plane of the bearing plate could be adjusted to make the driving spindle run vertically.

The coil armature plate was made from the same material as the baseboard but cut slightly larger than the section of the cage so that it could be secured by four 0B.A. bolts screwed down into each of the cage uprights. The coil armature was made from 6 mm mild steel rod bent into a 'U' shape in a vice so that the two ends had a similar spacing to that of the pole pieces of the magnet.

The armature plate was drilled to accommodate the armature. This was made by winding many turns of cotton-insulated thin copper wire on both legs of the armature. The coil was confined axially using sticky tape and longitudinally by 2 x 2 cm diameter tufnol washers on each leg. The coils were wound in opposite directions to produce a unidirectional current in them. The driving spindle was equipped with an over-bored *Meccano* pulley for flexible attachment to the universal motor (Pixii's generator of course used a handle and set-up gearing).

PERFORMANCE

The performance of the Pixii generator is shown in the table and the graph on the following pages. The O.C. e.m.f. obtained was 0.25 volt, and a current of 24 mA into a 1Ω load was obtained.

HIPPOLYTE PIXII MACHINE CURRENT INTO LOAD

CURRENT mA	LOAD (ohms)
2·0	110
2·3	90
3·1	70
4·3	50
5·1	40
6·2	30
8·3	20
10·2	15
12·7	10
15·4	7
17·3	5
20·4	3
22·1	2
24·0	1
Voltage, no load in circuit,	= 0·25 volt

TABLE 10 HIPPOLYTE PIXII GENERATOR CURRENT VS LOAD

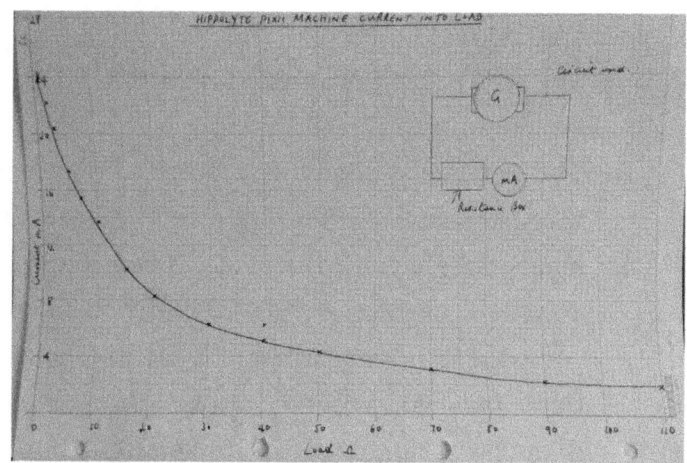

FIG. 42 HIPPOLYTE PIXII GENERATOR GRAPH OF CURRENT VS LOAD

APPENDIX H

TECHNICAL APPRAISAL OF THE SEPARATELY EXCITED GENERATOR

The separately excited generator is connected to an external source to energise it field magnets and the magnetic flux is therefore independent of the armature current except for the demagnetising component of the armature's field. The armature field induced by the current through it is in direction at right angles to the main field, but at the time of their introduction prior to the 1880's a commutator was fitted to produce D.C. and the armature field had the effect of altering the direction of the resultant field. This altered the angle of the armature at which no e.m.f. was induced and therefore that of the commutation. The brushes therefore had to advanced until the armature field was at right angles to the *resultant* field which produced a demagnetising competent on the main field :

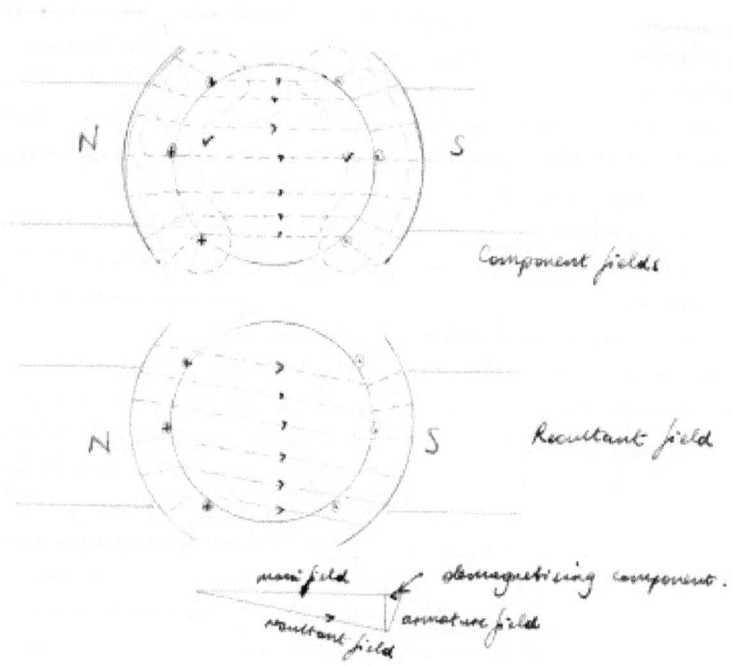

FIG. 43 DEMAGNETISING COMPONENT ON THE ARMATURE MAGNETIC FIELD DUE TO COMMUTATION

At a constant speed, the e.m.f. will be a constant independent of the load current. However, the e.m.f. has to overcome the resistance of the armature windings R_a and the voltage drop across the brushes V_b. The P.D. at the machine terminals is therefore:

$$V = E - I_a R_a - V_b,$$

where E is the e.m.f. of the machine and I_a the load current. A plot of V vs I_a is therefore virtually a straight line apart from variations in V_b where it has a negative slope:

FIG. 44 GRAPH OF OUTPUT CHARACTERISTIC OF THE SEPARATELY-EXCITED GENERATOR

TECHNICAL APPRAISAL OF THE SHUNT WOUND GENERATOR

The shunt wound self-excited generator has its field windings connected in parallel with the armature windings:

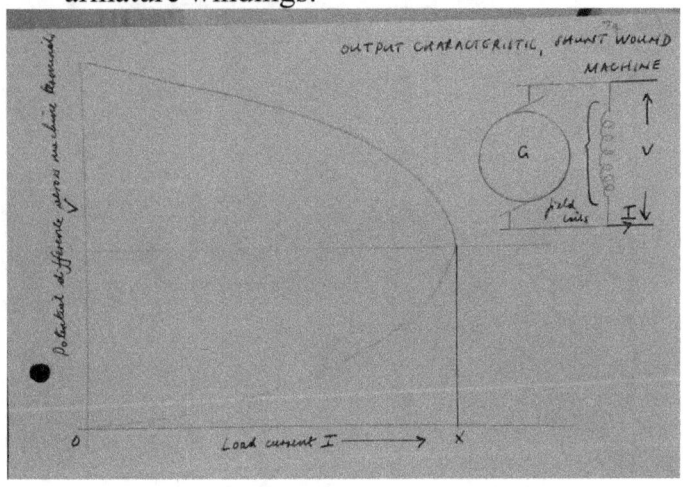

FIG. 45 GRAPH OF OUTPUT CHARACTERISTIC OF A SHUNT WOUND GENERATOR

When the machine is started on O.C., the residual magnetism in the field magnets starts the generation of an e.m.f. in the armature coils which increases the current in the field windings thus increasing the flux of the field. As the e.m.f. is proportional to the flux of the field magnets, the e.m.f. builds up. The magnetising field H of the field coils is proportional to the current in them, I_f, which on O.C. is equal to the armature current I_n. The field B in the armature is proportional to the e.m.f. of the machine E, and therefore the B-H curve is identical to the E vs I_a curve with suitable adjustment in scale.

The e.m.f. of the machine has therefore to overcome the potential drop in the armature and field windings,

$$I_f(R_a + R_f) + V_b.$$

When a load is connected across the machine terminals, part of the armature current passes through it. The armature current increases to compensate and therefore the potential drop $I_a R_a$ increases. The voltage across the machine's terminals $E - I_a R_a - V_b$ therefore decreases which in turn lowers I_f. The magnetising field therefore falls and so E and V decrease further. This state of affairs continues until E and V have values less than on O.C. The external characteristic is shown in Fig. 44. As the load increases, I_a increases and V falls. Therefore, the graph shows a falling characteristic As the load current becomes very large, the terminal voltage falls off rapidly and the e.m.f. produced in the machine becomes inadequate to supply the necessary field at a certain value of load X, and the machine shuts down.

TECHNICAL APPRAISAL OF THE SERIES AND COMPOUND WOUND MACHINES

The output characterise for the series wound machine is shown in the graph below:

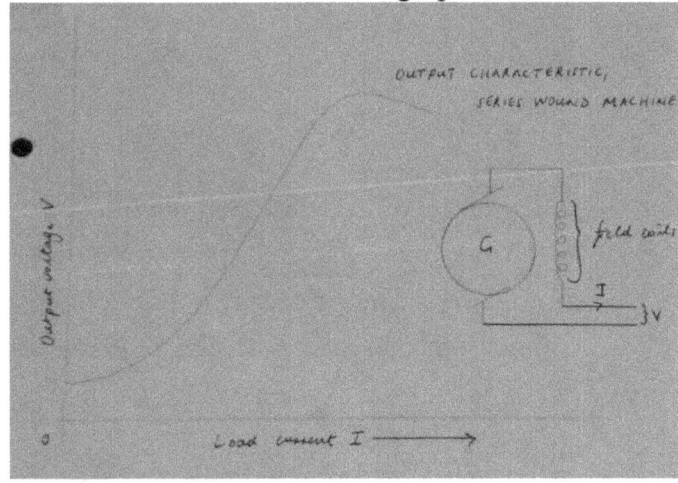

FIG. 46 GRAPH OF OUTPUT CHARACTERISTICS OF SERIES WOUND GENERATOR

In this case, the external current passes through the field coils which are connected in series with the armature windings, although the current in the field coils can be adjusted by means of a rheostat connected in parallel with the field windings. The characteristic curve is therefore the same shape as the B-H curve. The e.m.f. induced has to overcome the potential drop across the armature and field windings, and as the e.m.f. flattens off at the top of the magnetisation curve, the terminal voltage reaches a maximum then falls off.

Unlike the shunt wound a machine which has a falling characteristic, the e.m.f. and terminal voltage increase with load current in the series wound machine. The philosophy of Crompton was to combine series and shunt winnings in the field magnet circuit of self-excited machines to produce a much steadier output voltage which would be independent of load current over a broad range of loads. Machines equipped with Crompton's compound windings in the field magnet circuit have a characteristic of the shape shown below:

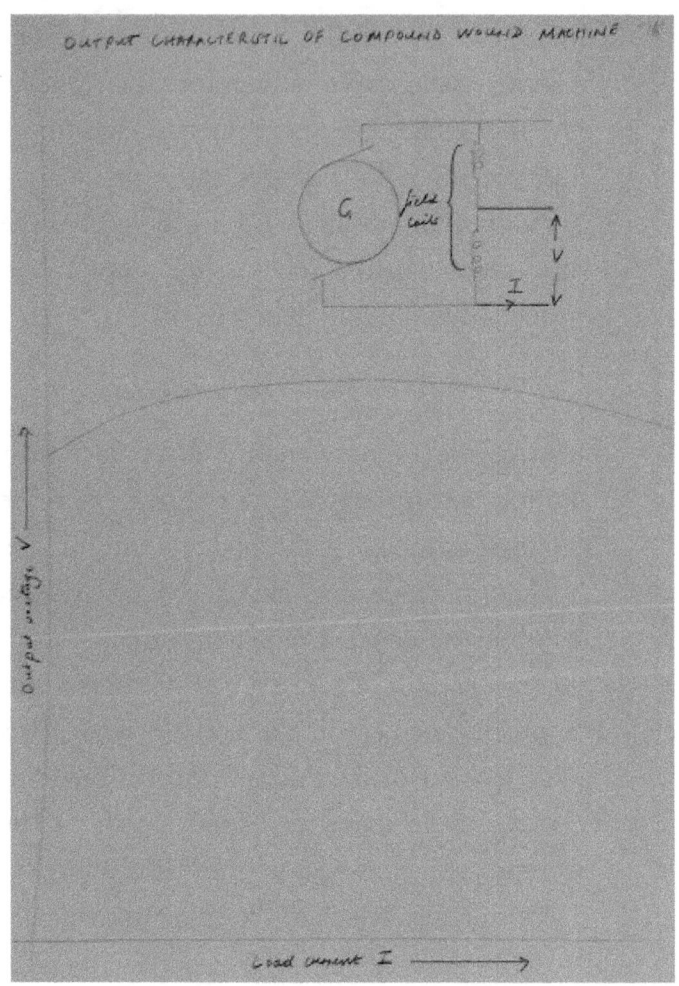

FIG. 47 OUTPUT CHARACTERISTRIC OF COMPOUND WOUND GENERATOR

APPENDIX I

VOLTAGE DIFFERENCES BETWEEN GENERATOR COILS

The figure below shows the output waveforms for each set of the three sets of coils in a 3-phase A.C. generator:

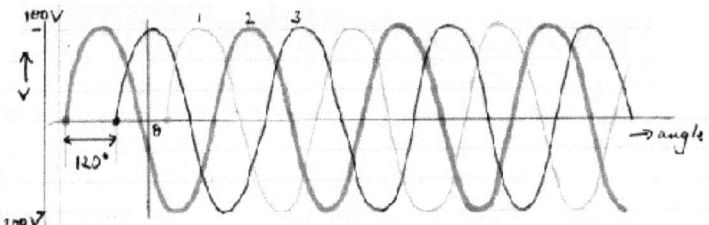

FIG. 48 THE THREE PHASE VOLTAGES IN A THREE-PHASE GENERATOR

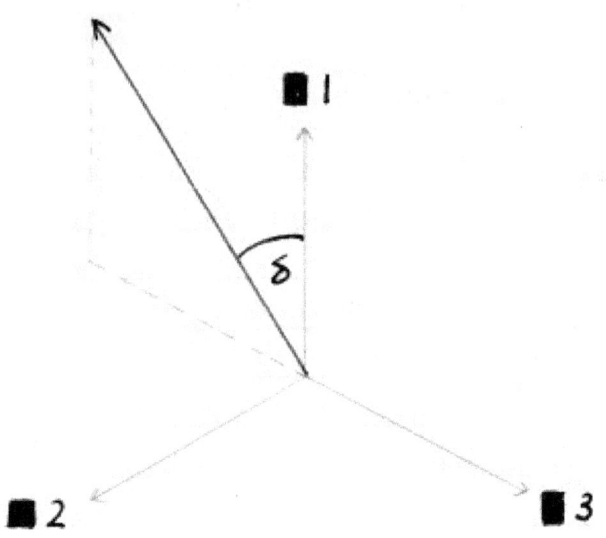

FIG. 49 VECTORIAL REPRESENTATION OF THE THREE PHASE VOLTAGES IN A THREE PHASE SYSTEM

In the diagram above, the voltage in 3, 2 or 1, the phase voltage, has magnitude and direction and so may be represented vectorially. The voltage across any two wires and the phase angle may be determined by summing the appropriate vectors. The P.D. between 3 and 1 is $V_1 - V_3$ and is shown by the black ink line in the vector diagram above. The magnitude of this vector is given by the cosine rule:

$$|V_1 - V_3|^2 = V_1^2 + V_1^2 - 2V_1 \cdot V_2 \cdot \cos 120°$$

$$= 2V_1^2 + 2V_1^2 \cdot 1/2$$

$$= 3V_1^2.$$

The P.D. between wires 1 and 3 is therefore $\sqrt{3}$ times the phase voltage and the angle δ

by which the voltage leads that in coil 1 is equal to 30°. Summation of vectors V_1 and V_2 and V_1 and V_3 give resultant

voltages equal to $\sqrt{3}$ times the phase voltage and lagging v_1 by 210 and 90 respectively.

ABOUT THE AUTHOR

Initially educated at a secondary modern school, leaving with 3 'A' levels Peta Trigger went on to university to gain honours degrees in Mathematics, Education and Engineering Science and a post graduate degree in Educational Psychology. Peta continued her education at London University's Institute of Education where she eventually completed her Ph.D and Ed.D degrees in Mathematics and Engineering Science Education. Her research included an investigation into landmark devices in the history of electricity generation and transmission.

She has produced articles on mathematics and science education, and previous work includes the book, 'Wind Turbines: Description, Appraisal & Alternatives'.

She is now 70 and has lived in Northampton for 23 years

www.ingramcontent.com/pod-product-compliance
Lightning Source LLC
Chambersburg PA
CBHW051632170526
45167CB00001B/156